Holden

Understanding Thermoplastic Elastome

Hanser **Understanding** Books

A Series of Mini-Tutorials

Series Editor: E.H. Immergut

Understanding Injection Molding Technology (Rees)
Understanding Polymer Morphology (Woodward)
Understanding Product Design for Injection Molding (Rees)
Understanding Design of Experiments (Del Vecchio)
Understanding Plastics Packaging Technology (Selke)
Understanding Compounding (Wildi/Maier)
Understanding Extrusion (Rauwendaal)
Understanding Thermoforming (Throne)
Understanding Thermoplastic Elastomers (Holden)

Geoffrey Holden

Understanding Thermoplastic Elastomers

Hanser Publishers, Munich

Hanser Gardner Publications, Inc., Cincinnati

The Author:
Dr. Geoffrey Holden, Holden Polymer Consulting, Incorporated
1042 Willow Creek Rd A111 – 273, Prescott, AZ 86301-1670, USA

Distributed in the USA and in Canada by
Hanser Gardner Publications, Inc.
6915 Valley Avenue, Cincinnati, Ohio 45244-3029, USA
Fax: (513) 527-8950
Phone: (513) 527-8977 or 1-800-950-8977
Internet: http://www.hansergardner.com

Distributed in all other countries by
Carl Hanser Verlag
Postfach 86 04 20, 81631 München, Germany
Fax: +49 (89) 98 12 64
http://www.hanser.de

Library of Congress Cataloging-in-Publication Data

Holden, G.
Understanding thermoplastic elastomers / Geoffrey Holden.
p. cm. – (Hanser understanding books)
Includes bibliographical references and index.
ISBN 1-56990-289-5
1. Elastomers. 2. Thermoplastics. 3. Copolymers. I. Title. II. Series.

TS1928.H65 2000
678–dc21 99-047357

Die Deutsche Bibliothek – CIP-Einheitsaufnahme

Holden, Geoffrey:
Understanding thermoplastic elastomers/ Geoffrey Holden. –
Munich: Hanser; Cincinnati: Hanser Gardner, 2000
(Hanser understanding books)
ISBN 3-446-19332-4

Typeset in the U.K. by Marksbury Multimedia Ltd, Bath
Printed and bound in Germany by Druckhaus Thomas Müntzer, Bad Langensalza

Contents

Contents

1 Introduction and Historical Survey

1.1 Overview

Thermoplastic elastomers are a relatively new development in the rubber industry. If we could step back in time to about 1960, we would find almost all the conventional (i.e., vulcanizable) rubbers that we are familiar with today being sold and used. The only significant exception would be hydrogenated nitrile rubber (HNBR). However, while thermoplastic polyurethanes had just been introduced, in 1960 all the other types of thermoplastic elastomers were yet to be discovered. Since then, the rapid growth of thermoplastic elastomers that indicates that there was clearly an unmet need for these products. Their worldwide annual consumption was estimated at 1,000,000 metric tons/year in 1995 [1]. This is expected to rise to about 1,400,000 metric tons/year in 2000, which amounts to a 7% annual growth rate.

Three books cover thermoplastic elastomers in detail. The first two concentrate mostly on the scientific aspects of these polymers [2,3] while the other concentrates on their end uses [4]. Thermoplastic elastomers have also been the subject of recent articles in encyclopedias and reference books [5–10]. The object of this book is to provide a short introduction and overview of the whole field, with the caveat that it is rapidly changing, so all that can be done in this text is to give a "snapshot" of the situation at the time of writing.

The outstanding advantage of thermoplastic elastomers can be summarized in a single phrase: they allow rubberlike articles to be produced by the rapid processing techniques developed by the thermoplastics industry. Their relationship to other polymers is shown in Table 1.1.

This table classifies all polymers by two characteristics: how they are processed (as thermosets or as thermoplastics) and the physical properties (rigid, flexible, or rubbery) of the final product. All the commercial polymers used for molding, extrusion, etc., fit into one of the six resulting classifications; the thermoplastic elastomers are the newest. They have many of the physical properties of rubbers, e.g., softness, flexibility, and

Table 1.1 Comparison of Thermoplastic Elastomers with Conventional Plastics and Rubbers

	Thermosetting	Thermoplastic
Rigid	Epoxies Phenol-Formaldehyde Urea-Formaldehyde	Polystyrene Polypropylene Poly(Vinyl Chloride) High Density Polyethylene
Flexible	Highly filled and/or highly vulcanized rubbers	Low Density Polyethylene Poly (Ethylene-Vinyl Acetate) Plasticized Poly(Vinyl Chloride)
Rubbery	Vulcanized Rubbers (Natural Rubber, Styrene-Butadiene Rubber, etc.)	Thermoplastic Elastomers

resilience. However, they achieve their properties by a physical process (solidification) compared with a chemical process (crosslinking) in vulcanized rubbers.

This solidification is accomplished in one of two ways: cooling or solvent evaporation, if the particular thermoplastic elastomer is soluble (most are not). In Table 1.2, the processes are compared with the vulcanization process needed to prepare usable articles from conventional rubbers.

In the terminology of the plastics industry, vulcanization is a thermosetting process. In other words, it is slow, irreversible, and usually requires heating. With thermoplastic elastomers, on the other hand, the transition from a processable melt to a solid, rubberlike, object is rapid, reversible, and takes place upon cooling. Thus, thermoplastic elastomers can be processed using conventional thermoplastic processing techniques, such as injection molding and extrusion. As with other thermoplastics, scrap can be recycled. Also, some thermoplastic elastomers are soluble in common solvents and can be processed as solutions.

Because they become soft and flow when heated, the high temperature properties of thermoplastic elastomers are usually inferior to those of conventional vulcanized rubbers. Thus, thermoplastic elastomers are usually not used in applications such as automobile tires. Instead, most of their applications are in areas where high temperature properties are less important, e.g., footwear, molded parts (including those used on automobiles), wire insulation, adhesives, and polymer blending.

Table 1.2 Transitions

Thermoplastic
Elastomers

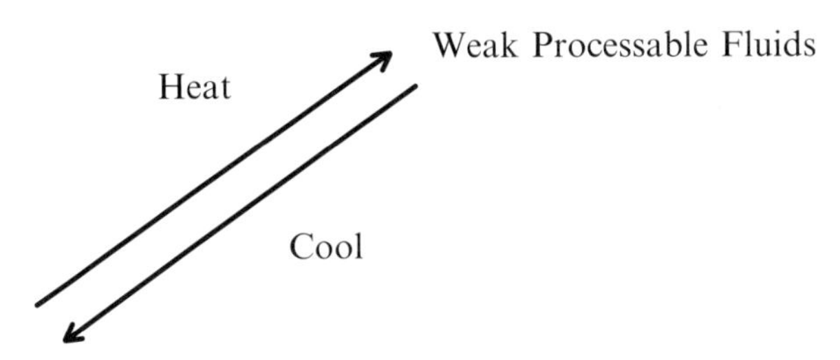

Strong Elastic Solids

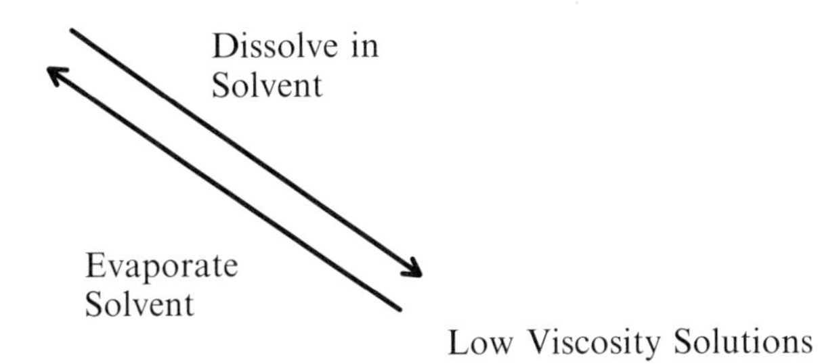

Conventional
Vulcanizates

Weak Processable Fluids —Heat, Time→ Strong Elastic Solids

1.2 Historical Survey

The history of thermoplastic elastomers is inevitably part of the development of the whole history of polymer chemistry [11]. Although natural high polymers (e.g., cotton, wool, rubber) have been known for centuries, there was no understanding of the nature of these materials; essentially people used what was available. Probably the first significant attempt to improve on nature was the crosslinking (or vulcanization) of rubber, developed by Charles Goodyear in 1839. A few years before this, John Hancock reduced the molecular weight of rubber by milling. These two discoveries became the foundation of the rubber industry.

Note that both these discoveries were made pragmatically. There was no understanding of the principles involved. Williams and later, Harries, showed that natural rubber was derived from isoprene (C_5H_8). However, it was believed that rubber was some kind of aggregation or colloidal association, possibly of low molecular weight cyclic structures based on isoprene, what today would be called cyclic oligamers. Similar structures were proposed for proteins and cellulose. These and similar materials were considered low molecular weight compounds, held together by secondary valance forces. Very large molecules were thought to be impossible. Indeed, Emil Fischer, the famous organic chemist and Nobel Laureate, maintained that 5,000 was the upper molecular weight limit for organic compounds.

Despite this lack of fundamental understanding, important discoveries continued to be made. About 1870, Celluloid (a mixture of cellulose nitrate and camphor) was introduced. In 1910, Baekeland developed the first synthetic resin, Bakelite, derived from phenol and formaldehyde. In Germany during World War I, dimethyl butadiene was polymerized to produce a substitute for natural rubber.

The years before World War II saw two parallel developments. One was the introduction of more synthetic polymers; polystyrene (PS), polyvinyl chloride (PVC), and styrene butadiene rubber (SBR) are outstanding examples. The second was the development of the fundamental theory of high polymers by Staudinger and later, Carothers [11]. For the first time, workers in this field understood what was happening during polymer manufacture and processing. Carothers used this understanding to develop nylon and neoprene, the first a thermoplastic, and the second an elastomer.

The first developments in thermoplastic elastomers also occurred about this time [12]. The first work was based on PVC, a rigid thermoplastic. It contains a significant amount of syndiotactic structure that can crystallize and also amorphous atactic structure. (Tacticity is described in more detail in Section 4.2). At room temperature, the syndiotactic structure is crystalline, and the amorphous atactic structure is above its glass transition temperature. Thus, both structures are hard and rigid at room temperature. However, as Semon discovered about 1930, plasticizers can be added (e.g., dioctyl phthalate (DOP)) that swell the atactic polymer and reduce its glass transition temperature to well below room temperature. This converts it to a flexible product. The result is what we now know as the structure of most thermoplastic elastomers: a combination of a rigid phase (syndiotactic PVC) that becomes fluid at processing temperatures with a softer, flexible phase (plasticized atactic PVC).

However, plasticized PVC is not usually considered an elastomer. It lacks many elastomeric properties such as snapback, resilience, and high surface friction. But it was the first material that even came close to being a thermoplastic elastomer. In 1940, its elastic properties were improved by blending with another elastomer, nitrile rubber (NBR). PVC / NBR / DOP blends are now an important part of the thermoplastic rubber industry.

About 1937, workers in Germany at I. G. Farben developed the urethane reaction between an isocyanate and an alcohol. By using diisocyanates and glycols, the result was a long chain structure, similar in principle to nylon. By using two glycols (one short chain, the other long), blocks of two polyurethanes are produced. The first is crystalline; the second amorphous. Again, they form the basic two-phase system characteristic of most thermoplastic elastomers. Starting about 1955, this principle was used by workers at DuPont and at B. F Goodrich to produce elastic fibers and moldable rubbers. It was later extended to yield thermoplastic elastomers with both polyester and polyamide hard segments.

In the 1950s, anionic polymerization was developed. In this system, solution polymerization is initiated by a metallic anion, (e.g., sodium). Pure metals were used at first, but alkyl-metallics (e.g., butyl lithium) were found to give better results. The system is "living"; that is, in the absence of terminating agents, the polymeric product can initiate further polymerization. Thus, if a second monomer is added, the result is a block copolymer. Styrene, butadiene, and isoprene are the only common monomers that can easily be polymerized in this way. The first commercial products were polybutadiene and polyisoprene (anionic production of polystyrene is not economic). In 1961, attempts at Shell Chemical to improve the cold flow properties of these two elastomers led to the development of styrenic block copolymers. These were important for two reasons:

1. They offered a low-cost route to the production of thermoplastic elastomers with many properties of conventional vulcanized rubbers.
2. Their simple and unequivocal structure gave a clear picture of how other thermoplastic elastomers (or at least, those based on block copolymers) gained their properties. In other words, they served as model polymers.

Later, about 1975, similar polymers with improved stability were produced by selective hydrogenation of the elastomer segments in these block copolymers.

In the 1960s, other block copolymers, such as polycarbonate/polyether and poly (silphenylene siloxane)/polydimethylsiloxane, were found to have elastomeric properties without vulcanization, but the reasons were not clearly understood. Many other hard polymer/elastomer block copolymers have been investigated since then and have been shown to produce thermoplastic elastomers. For example, carbo cationic polymerization has been used to produce thermoplastic elastomers from block copolymers of styrene and isobutylene (see section 3.3), and metallone catalysts have been used to produce thermoplastic elastomers from block copolymers of olefins.

The basic requirements for a thermoplastic elastomer, a hard phase and an elastomeric phase, were now established. As well as hard polymer/ elastomer block copolymers, there are several other ways of achieving this requirement. An obvious one is simple mixing. About 1960, two new polymers (both produced by Ziegler-Natta catalysts) were introduced. The first was a rubber, a copolymer of ethylene and propylene (EPR). The second was a thermoplastic, isotactic polypropylene. They are produced from low-cost monomers and should obviously be technically compatible with each other. At first only a small amount of EPR was mixed with the polypropylene to produce a high impact thermoplastic. When more EPR was added, often extended with oil, the result was a hard thermoplastic elastomer. Attempts to produce softer grades in this way were less successful; the large proportion of the weak EPR phase resulted in poor properties.

However, this problem was solved around 1975, when the elastomer phase (in this case, EPDM) was crosslinked during the mixing process in a system called "dynamic vulcanization". The resulting thermoplastic elastomers can be quite soft and their properties are often better than those of simple mixtures.

Other systems investigated include graft copolymers (an elastomer chain on which is grafted several hard segments) and elastomeric ionomers (an elastomer chain containing acidic groups with associated metal cations). While they have many interesting properties, they have not developed into commercial products.

References

1 Reisch, M.S., (1996) *Chemical and Engineering News, 74(32)*, 10.
2 Legge, N.R., Holden G. and Schroeder, H.E., Eds. *Thermoplastic Elastomers - A Comprehensive Review* (1987) Hanser Publishers, Munich, Vienna, New York.
3 Holden, G., Legge, N.R., Quirk, R.P. and Schroeder, H.E., Eds., *Thermoplastic Elastomers*, 2nd Ed., (1996), Hanser Publishers, Munich, Vienna, New York.
4 Walker, B.M. and Rader, C.P., Eds., *Handbook of Thermoplastic Elastomers*, 2nd Ed., (1998), Van Nostrand Reinhold, New York
5 Holden, G., "Thermoplastic Elastomers" in *Rubber Technology*, 3rd Ed., M. Morton, Ed., (1987), Van Nostrand Reinhold, New York
6 Holden, G., "Elastomers, Thermoplastic" in *Kirk-Othmer Encyclopedia of Polymer Science and Engineering*, 2nd Ed, J. I. Kroschwitz, Ed., (1986), John Wiley & Sons, Inc., New York

7 Holden, G., "Elastomers, Thermoplastic" in *Kirk-Othmer Encyclopedia of Chemical Technology*, 4th ed, J. I. Kroschwitz, Ed., (1988) John Wiley & Sons, New York.
8 Holden, G., "Thermoplastic Elastomers (Overview)" in *Polymeric Materials Encyclopedia*, J. C. Salamone, Ed. (1996), CRC Press, Boca Raton
9 Holden, G. and Wilder, C.R., "Thermoplastic Styrenic Block Copolymers" in *Handbook of Elastomers*, 2nd Ed., E. K. Bhowmick and H. L. Stevens, Eds., (In Press), Marcel Dekker, Inc., New York Basel
10 Holden, G., "Thermoplastic Elastomers and Their Applications" in *Applied Polymer Science - 21st Century*, C. Carraher and C. Craver, Eds., (In Press), American Chemical Society. Washington
11 Furukawa, Y., *Inventing Polymer Science*. (1998), University of Pennsylvania Press, Philadelphia
12 Holden, G., Legge, N.R., Quirk, R.P. and Schroeder, H.E. in Chapter 1 of Ref. 1

2 Classification and Structure

Thermoplastic elastomers can be classified into seven groups. They are:

1. Styrenic block copolymers
2. Crystalline multi-block copolymers
3. Miscellaneous block copolymers
4. Hard polymer/elastomer combinations
5. Hard polymer/elastomer graft copolymers
6. Ionomers
7. Polymers with core-shell morphologies

Many of the styrenic block copolymers, the crystalline multi-block copolymers, and the hard polymer/elastomer combinations are commercially important. In addition, there is at least one other type of block copolymer that is a commercial product. The commercial uses of these polymers are covered in Chapter 7.

With such a variety of materials, it is to be expected that the properties of thermoplastic elastomers cover an exceptionally wide range. Some are very soft and rubbery while others are hard and tough, and in fact approach the ill-defined interface between elastomers and flexible thermoplastics.

2.1 Phase Structure

Most thermoplastic elastomers have one feature in common; they are phase-separated systems The Alcryn* thermoplastic rubbers and those based on ionomers are possible exceptions. One (sometimes more than one) phase is hard and solid at room temperature while another phase is an elastomer and fluid. Often the phases are chemically bonded by block or graft copolymerization. In other cases, a fine dispersion of the phases is apparently sufficient. The hard phase gives these thermoplastic elastomers

*Alcryn is a registered trademark of Advanced Polymer Alloys

their strength. Without it, the elastomer phase would be free to flow under stress and the polymers would be unusable. When the hard phase is melted, or dissolved in a solvent, flow can take place and so the thermoplastic elastomer can be processed. On cooling or evaporation of the solvent, the hard phase solidifies and the thermoplastic elastomers regain their strength. Thus, in a sense, the hard phase in a thermoplastic elastomer acts similarly to the sulfur crosslinks in conventional vulcanized rubbers and the process by which it does so is often called physical crosslinking. Conversely, the elastomer phase provides elasticity and flexibility to the system.

2.2 Phase Properties

Because most thermoplastic elastomers are phase-separated systems, the individual polymers that constitute the phases retain many of their characteristics despite being combined together. For example, each phase has its own crystal melting point (T_m) and/or glass transition temperature (T_g). These determine the temperatures at which a particular thermoplastic elastomer goes through transitions in its physical properties. Thus, when the properties (e.g., modulus) of a thermoplastic elastomer are measured over a range of temperatures, there are three distinct regions (see Fig. 2.1). At very low temperatures, both phases are hard and so the material is stiff and brittle. At a somewhat higher temperature the elastomer phase becomes

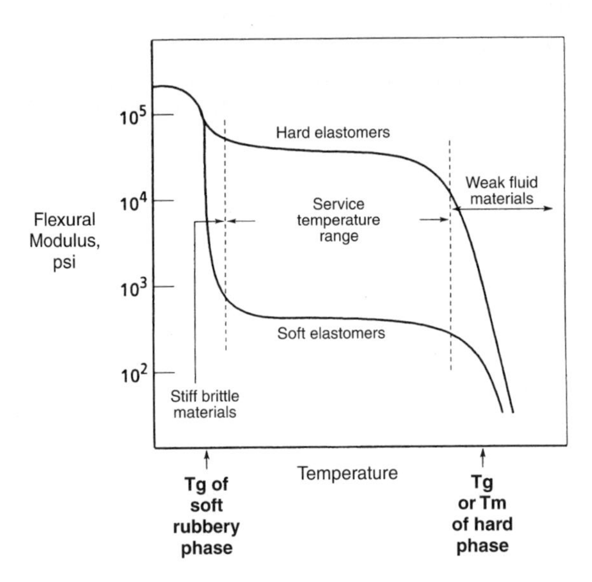

Figure 2.1 Stiffness of typical thermoplastic elastomers at various temperatures

soft and the thermoplastic elastomer now resembles a conventional vulcanizate. As the temperature is further increased, the modulus stays relatively constant (a region often described as the "rubbery plateau") until finally, the hard phase softens. At this point, the thermoplastic elastomer becomes fluid. Thus, thermoplastic elastomers have two service temperatures. The lower service temperature depends on the T_g of the elastomer phase while the upper service temperature depends on the T_g or T_m of the hard phase. The difference between the upper and lower service temperatures is the service temperature range.

Values of T_g and T_m for the various phases in some commercially important thermoplastic elastomers are given in Table 2.1. Note that because the elastomer phase begins to harden above its T_g, practical lower service temperatures are somewhat higher than the values given in this table. The exact values depend on the hardening that can be tolerated in the

Table 2.1 Glass Transition and Crystal Melting Temperatures[a]

Thermoplastic Elastomer Type	Soft, Rubbery Phase Tg (°C)	Hard Phase Tg or Tm (°C)
Polystyrene/Elastomer Block Copolymers		
S-B-S	−90	95(T_g)
S-I-S	−60	95(T_g)
S-EB-S	−60	95(T_g) & 165(T_m)[b]
S-iB-S	−60	95–240[c]
Multi-block copolymers		
Polyurethane/Elastomer Block Copolymers	−40 to −60[d]	190(T_m)
Polyester/Elastomer Block Copolymers	−40	185 to 220(T_m)
Polyamide/Elastomer Block Copolymers	−40 to −60[d]	220 to 275(T_m)
Polyethylene/Poly(α-olefin) Block Copolymers	−50	70(T_m)
Hard Polymer/Elastomer Combinations		
Polypropylene/EPDM or EPR combinations	−60	165(T_m)
Polypropylene/Nitrile Rubber combinations	−40	165(T_m)
PVC/Nitrile Rubber/DOP combinations	−30	80(T_g)

[a]Measured by Differential Scanning Calorimetry
[b]In compounds containing polypropylene
[c]The higher values are for substituted polystyrenes and polyaromatics (see Table 3.4)
[d]The values are for polyethers and polyesters, respectively

final product. Similarly, the hard phase in the thermoplastic elastomer begins to soften below its T_g or T_m, and so practical upper service temperatures are somewhat lower than the values given in Table 2.I. The upper service temperatures also depend on the stress applied. An unstressed part (e.g., one undergoing heat sterilization) has a higher upper service temperature than one that must support a load.

The choice of polymer in the hard phase strongly influences the oil and solvent resistance of the thermoplastic elastomers. Even if the elastomer phase is resistant to a particular oil or solvent, if this oil or solvent swells the hard phase, all the useful physical properties of the thermoplastic elastomer are lost. In most thermoplastic elastomers, this hard phase is crystalline and so it is resistant to oils and solvents. Styrenic thermoplastic elastomers are an exception. As pure polymers, they are soluble in common solvents and so, have poor oil and solvent resistance, although this can be improved by compounding with crystalline polymers. However, this solubility means they can be applied from solution, an advantage over many other thermoplastic elastomers.

In addition to all the possible components for the two phases and the ways they can be combined, the relative amounts of the hard and the soft phases can be varied, as well. As might be expected, increasing the ratio of hard to soft phases makes for a stiffer product. All these variations in type, structural arrangement, and relative proportion of the hard and soft phases lead to an exceptionally wide range of properties and applications for thermoplastic elastomers. It is the aim of this book to cover them in sufficient detail to provide the reader with a working knowledge of these materials.

2.3 Polymer Structure

Although the first three classes of thermoplastic elastomers listed above are all block copolymers, they have important structural differences. Most of the styrenic block copolymers have the general formula S-*E*-S, where S represents a hard polystyrene segment and *E* an elastomeric segment. Other styrenic block copolymers have a branched structure with the general formula $(S\text{-}E)_n x$, where x represents an n-functional junction point.

In contrast, multi-block copolymers with crystalline hard segments have the general formula *H-E-H-E-H-E...* or $(H\text{-}E)_n$, where *H* represents a hard thermoplastic block that is crystalline at service temperatures. Typical examples are polyurethanes, polyesters, polyamides, and polyethylene. In these block copolymers, the molecular weight distributions of both the individual blocks and the polymer as a whole are very broad. The segmental molecular weights are relatively low compared with those of the styrenic block copolymers.

The third class of block copolymers consists of miscellaneous materials that are thermoplastic elastomers, but have few or no commercial applications. The hard polymer/elastomer combinations are intimate mixtures of the two phases. Sometimes the elastomer phase is crosslinked.

Thermoplastic elastomers have also been produced from graft copolymers. These may be represented as

$$E \ldots E\text{-}\left[\begin{array}{c}\text{-}E\text{-}\\ | \\ H\end{array}\right]_n\text{-}E\text{-}E\text{-}E \ldots E$$

This represents a polymer where each elastomeric *E* polymer chain has (on average) n random grafts of hard *H* blocks. Although much effort has been expended in research on these materials, they have not become commercially important.

There has been some work on thermoplastic elastomers produced from ionomers. These may be represented as:

$$\begin{array}{l}P \ldots P\text{-}P\text{-}P^{-}\ \text{-}P\text{-}P \ldots P\\ \qquad\quad (\mathrm{M}^{+})_n\end{array}$$

This represents a polymer where each backbone polymer *P* block has (on average) n acidic functionalities with associated metallic (M^+) counterions. The metallic counterions cluster together to give "ionic crosslinks" that are thermally labile. As originally developed, the backbone polymer was polyethylene. The products were clear, flexible thermoplastics and were commercialized by DuPont under the trade-name IONOMER. Later this work was extended to include other polymers such as sulfonated EPDM, a rubbery copolymer of ethylene, propylene, and small amount of a diene. It has reactive double bonds attached to the saturated elastomer chain. These materials yielded products with more elastomeric properties. Thermoplastic elastomers based on this concept have been produced by compounding these sulfonated polyethylenes with metal stearates, oils, fillers, and polyolefins, but so far they do not appear to have commercial applications.

2.3.1 Nomenclature

It is common practice to use the first letter of the monomer unit to denote the polymer block. For example, a three-block copolymer, poly(styrene-

block-butadiene-*block*-styrene) is written as S-B-S. Occasionally, two letters must be used to avoid confusion. Thus, the block copolymer poly(styrene-*block*-isobutylene-*block*-styrene) can be written as S-iB-S. If a block is itself a copolymer (e.g., ethylene-propylene rubber), the block copolymer poly(styrene-*block*-ethylene-*co*-propylene-*block*-styrene) is written as S-EP-S. These abbreviations are defined the first time they are used in this text. They are also listed in Table 8.6. There are many accepted abbreviations of common polymers (e.g., EPR for a random copolymer of ethylene and propylene). Those used in this book are also listed in Table 8.6.

3 Styrenic Block Copolymers

Styrenic block copolymers are based on simple molecules such as an S-*E*-S block copolymer, where S is a polystyrene segment and *E* is an elastomer segment. The two segment types are incompatible and so form a two-phase system. The most common are those for which the elastomer segment is a hydrocarbon. Others, e.g., those with a silicone rubber as the elastomer, are less important. Examples of all these are shown in Table 3.1.

Two basic polymerization systems – anionic [4] and carbocationic [5] are used to produce these styrenic block copolymers. However, the products share many common features that can be considered together.

3.1 Common Features

3.1.1 Domain Theory

If the elastomer is the major constituent, the block copolymers should have a morphology similar to that shown in Fig. 3.1. Here, the polystyrene end

Table 3.1 Thermoplastic Elastomers Based on Styrenic Block Copolymers

Hard Segment, *H*	Soft or Elastomeric Segment, *E*	Structure[a]	Refs.
Polystyrene	Polybutadiene and polyisoprene	T, B	[1–4]
Polystyrene	Poly(ethylene-co-butylene) and Poly(ethylene-co-propylene)	T	[2, 3]
Polystyrene & Substituted Polystyrenes	Polyisobutylene	T, B	[5]
Poly(α-methyl-styrene)	Polybutadiene, polyisoprene	T	[4]
Poly(α-methyl-styrene)	Poly(propylene sulfide)	T	[4]
Polystyrene	Polydimethylsiloxane	T, M	[6]
Poly(α-methyl-styrene)	Polydimethylsiloxane	M	[4, 6]

a. T = Triblock, *H-E-H*, B = Branched, $(H\text{-}E)_n$x, M = Multiblock, *H-E-H-E-H-* . . .

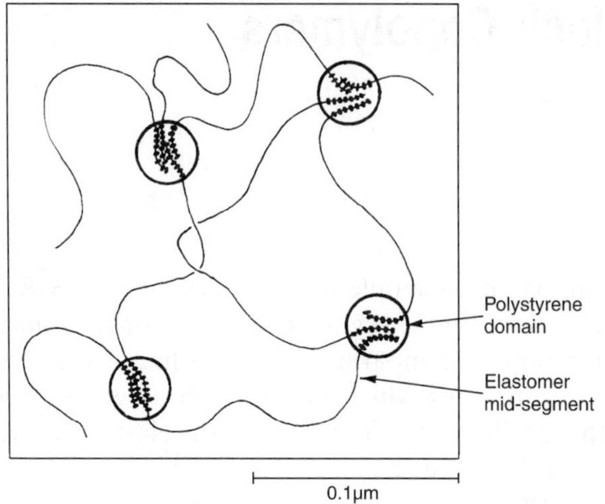

Figure 3.1 Morphology of styrenic block copolymers

segments form separate spherical regions, i.e., domains, dispersed in a continuous elastomer phase. Most of the polymer molecules have their polystyrene end segments in different domains. At room temperature, these polystyrene domains are hard and act as physical crosslinks, tying the elastomeric mid-segments together in a three-dimensional network. In some ways, this is similar to the network formed by vulcanizing conventional rubbers using sulfur crosslinks. The difference is that in thermoplastic elastomers, the domains lose their strength when the material is heated or dissolved in solvents. This allows the polymer or its solution to flow. When the material is cooled or the solvent is evaporated, the domains harden and the network regains its original integrity.

This domain theory explains what at first seemed almost unbelievable properties. The purpose of vulcanization is to give strength, extensibility, and "snap" to rubbers. For this reason, almost all rubber products in service are vulcanized (or at least they were at the time thermoplastic elastomers were discovered). Yet here was an *unvulcanized* rubber, based on styrene and butadiene monomers, that had better strength and extensibility than a *vulcanized* rubber prepared from SBR, a random copolymer of the same two monomers (see Fig. 3.2). In paper largely devoted to the physical properties of S-B-S and S-I-S block copolymers [1], the domain theory was first postulated in an attempt to explain this behavior. It was also extended to explain the viscous flow of these polymers. However, there was no direct evidence. Since then, many electron micrographs have been published, some showing a remarkably well defined and regular structure [7] (see Fig. 3.3).

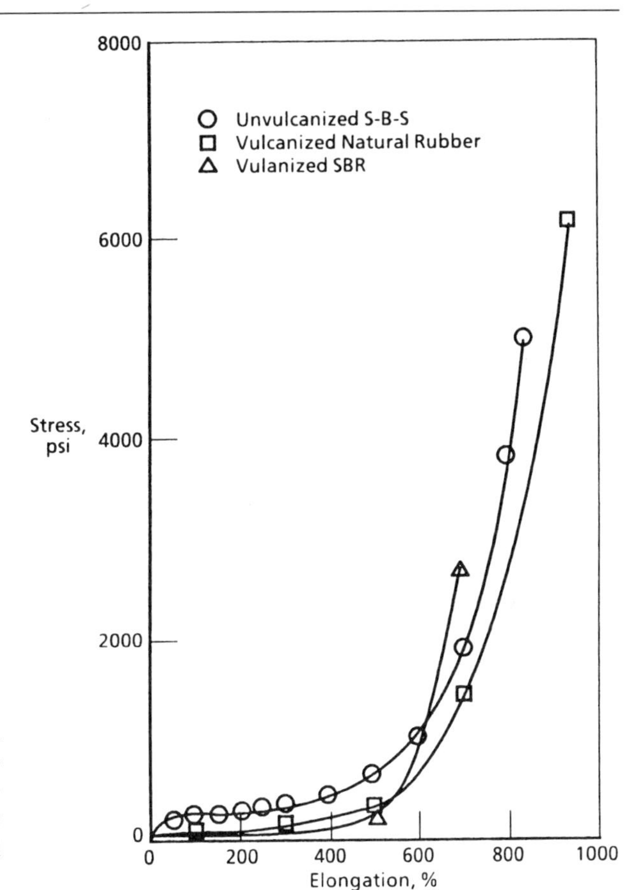

Figure 3.2 Stress–strain properties of an unvulcanized S-B-S block copolymer compared to two vulcanized conventional rubbers

These micrographs showed clear evidence of a two-phase system that (at lower polystyrene contents) is similar to that postulated in Fig. 3.1.

As will be discussed in Section 3.2.2, these elastomer chains are highly entangled. Many of their critical strength properties derive from these entanglements [1]. As well as functioning as crosslinks, the hard domains prevent these entanglements from disentangling under stress. It should be noted that analogous block copolymers with only one hard segment (e.g., S-*E* or *E*-S-*E*) have quite different properties [1]. In these polymers, the elastomer phase cannot form a continuous interlinked network because only one end of each elastomer segment is attached to the hard domains. These polymers are not thermoplastic elastomers, but are weaker materials similar to *unvulcanized* synthetic rubbers.

3.1.2 Filler Effects

In addition, the hard polystyrene domains act as reinforcing fillers. In conventional vulcanized rubbers, reinforcing fillers (e. g., carbon black and silica) are used to produce improved physical properties, particularly tensile and tear strength. They are particularly important in formulating rubbers that, because of their irregular structures, cannot stress-crystallize (e.g., SBR and EPR). Without reinforcing fillers, these rubbers have poor physical properties and so do not have many practical applications.

Effective reinforcing fillers such as carbon black are hard, small, well-dispersed particles that are strongly bonded (either physically or chemically) to the elastomer chains and prevent small stress-induced cracks in the elastomer from spreading and initiating catastrophic rupture. The hard polystyrene domains in S-*E*-S block copolymers fulfill all these requirements. They are small (typically less than 300 Angstroms in diameter, (see Fig. 3.3) and well dispersed. They are strongly bonded to the elastomer phase because they are part of the same molecule. In most cases, the elastomer segments in S-*E*-S block copolymers cannot stress-crystallize and so the role of the domains as reinforcing filler particles is crucial to their applications.

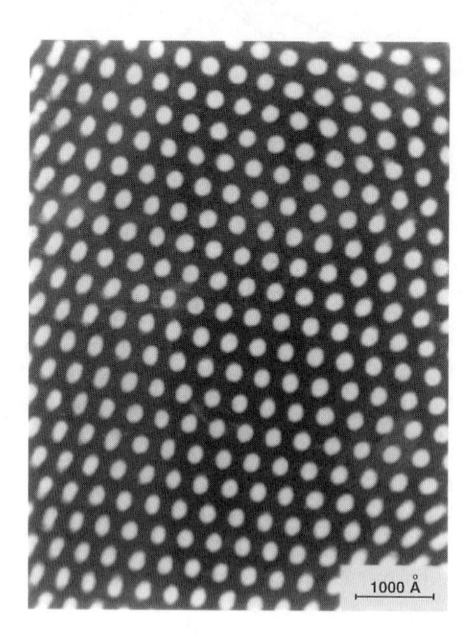

Figure 3.3 Electron micrograph of an $(S\text{-}I)_n$ block copolymer. The elastomer phase is stained black

Thus, the polystyrene domains in styrenic block copolymers fulfill three critically important functions:

1. They act as physically reversible crosslinks and tie the elastomer chains together to form a network that is effective at service temperatures but can flow at higher temperatures and disperse in suitable solvents.
2. They prevent the entangled elastomer chains in this network from disentangling under stress.
3. They act as reinforcing fillers.

3.1.3 Polystyrene/Elastomer Ratio

As in other thermoplastic elastomers, the morphology of these styrenic block copolymers depends on the ratio of the volume of the hard polystyrene phase to that of the softer elastomer phase. This ratio can be varied within quite wide limits. In the general case [8], in a block copolymer of *A and B*, as the ratio of the *A* to *B* segments is increased, the phase morphology changes from a dispersion of spheres of *A* in a continuous phase of *B* to a dispersion of rods of *A* in a continuous phase of *B*. Further increases in the ratio of the *A* to *B* segments yields a lamellar or "sandwich" structure in which both *A* and *B* are continuous. If the proportion of *A* is increased still further, the effect is reversed; *A* now becomes disperse and *B* continuous (see Fig. 3.4). A series of electron micrographs published shortly after the domain theory was developed [9] showed exactly this effect in four S-B-S block copolymers of increasing polystyrene contents (Fig. 3.5). At lower polystyrene levels, (Figs 3.5a and 3.5b), the micrographs showed a

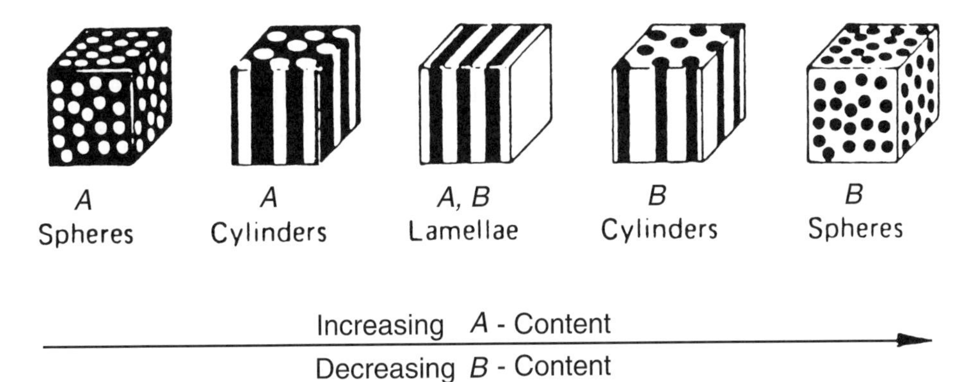

Figure 3.4 Morphology changes with composition in *A-B-A* block copolymers

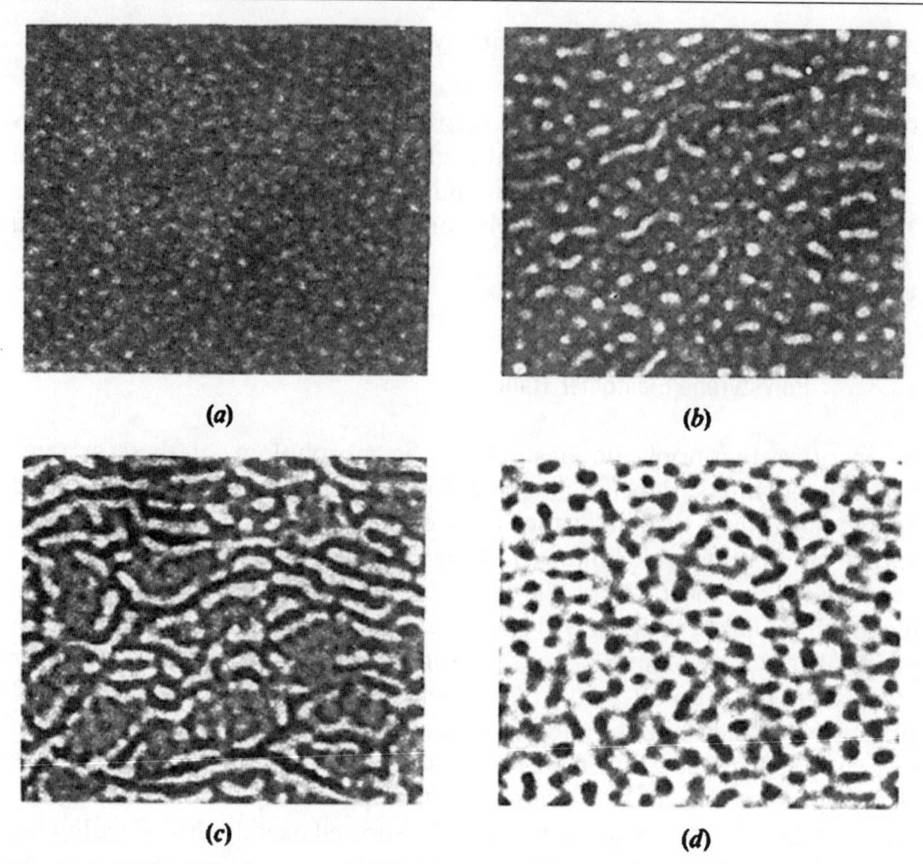

Figure 3.5 Electron micrograph of an S-B-S block copolymers. The elastomer phase is stained black

dispersion of polystyrene in a continuous phase of polybutadiene (stained black in the figures). As the polystyrene increased to 50% of the total volume, this structure changed to two co-continuous phases (Fig. 3.5c). As the polystyrene content increased still further to 84%, there was a phase inversion to a dispersion of polybutadiene in a continuous phase of polystyrene. Because of this change in morphology, as the polystyrene content increases, the stress-strain behavior changes [1] from soft, weak material to a strong elastomeric material and then, at about 45% polystyrene content, to a hard, leathery material with a yield point (see Fig. 3.6). As the polystyrene phase increases still further, the block copolymer gets harder and stiffer until eventually it becomes a clear, flexible thermoplastic (e.g., Phillips K-Resin).

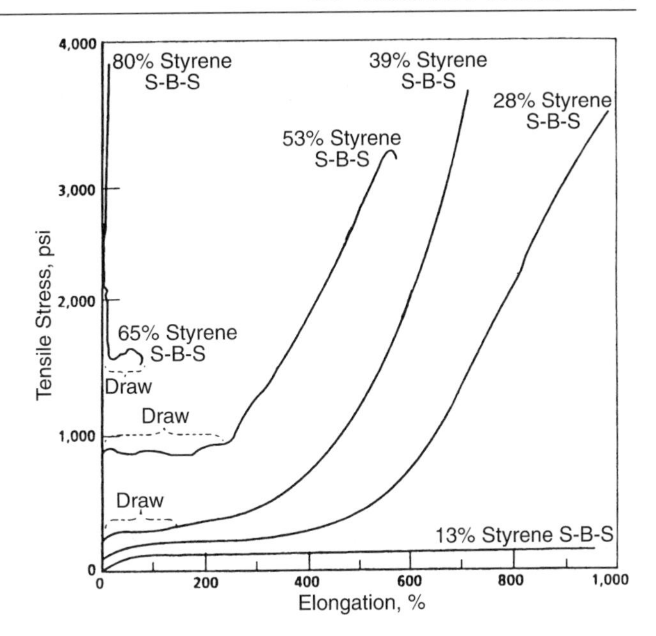

Figure 3.6 Stress–strain properties of styrenic block copolymers at varying polystyrene contents

3.1.4 Clarity

As is discussed in more detail in the next sections, block copolymers of reasonably high molecular weight form a phase separated system. Phase separation, with a few exceptions, is characteristic of mixtures of high molecular weight polymers. If the phases differ in refractive index, as they almost always do, two-phase systems are opaque. An everyday example of this is a dispersion of oil and water, which is opaque, although the oil and the water are both clear when viewed separately. Similarly, a mixture of polybutadiene and polystyrene is opaque, although both polybutadiene and the polystyrene are clear in the unmixed state. The reason for the opacity of the mixture is that light is diffracted ("bent") as it passes through the interface between the phases. Because the interface is usually not flat, it scatters the light randomly, and the mixture appears milky or white.

However, block copolymers such as S-B-S are quite clear because of the size of the phases [1]. If the molecular weight of the polystyrene segments is about 10,000, then the extended length of each polystyrene chain is about 200 Angstroms. Because the junction between the polystyrene segments and the polybutadiene segments must be at the surface of the polystyrene domains, the diameter of these domains (or their thickness if they are lamellae) cannot be greater than about 400 Angstroms. If it were greater, they would have holes in the middle! Thus, the maximum size of the polystyrene domains is much less than the wavelength of light (about 5,000 Angstroms). Because of this, the polystyrene domains do not scatter light significantly and S-B-S and similar block copolymers appear transparent.

The discussion above used a simple approach to estimate a maximum domain size. In practice, because of packing considerations, the observed domain size is somewhat less (see Fig. 3.3). The S-B-S block copolymer was the example used, but the same principle applies to other block copolymers. As a result, we can come to the generalized conclusion that pure block copolymers of all types, unless their molecular weights are much higher than the range of practical interest, should appear transparent, which in fact they are.

However, such an appearance does not occur in all thermoplastic elastomers. Those based on blends and thermoplastic vulcanizates have no restrictions on maximum phase size. Thus, they form much coarser dispersions, typically about 10,000 Angstroms, and so appear opaque.

3.1.5 Thermodynamics of Phase Separation

For the styrene domains to be effective, they must be able to separate from the elastomer phase. Generally, materials mix to form a single phase system if the free energy of mixing (ΔG_m) is favorable, i.e., negative. This free energy can be expressed as:

$$\Delta G_m = \Delta H_m - T\,\Delta S_m \tag{3.1}$$

where ΔH_m and ΔS_m are the enthalpy and entropy of mixing and T is the absolute temperature. Thus, the condition for domain formation is that:

$$\Delta H_m > T\,\Delta S_m \tag{3.2}$$

For hydrocarbon polymers, ΔH_m is usually positive because there are no strongly intercting groups. It increases as the structures of the two polymers forming the segments become less alike. ΔS_m always is positive and will depend on the number of segment that mix. Thus it approaches zero as the molecular weights of the segments become very large. Thus, we can expect domain formation to be favored by the following factors:

1. A high degree of structural difference between the segments
2. High segmental molecular weight
3. Low temperatures

Using this approach, the theory of domain formation has been extensively developed and quantified [10, 11]. In experimental work, the effects of structural differences are shown by the fact that E-EB-E block copolymers (where E represents a polyethylene segment) are apparently not phase separated in the melt [4] while a strong separation exists for corresponding S-EB-S block copolymers [12]. The effects of molecular weight and temperature have been demonstrated by work on an experiental S-B-S block copolymer with end segments having molecular weights of 7,000. Viscosity measurements [13] suggest that this polymer changes to a one-phase system at about 150 °C. On the other hand, similar block copolymers with end segment molecular weights of 10,000 and greater appear to be phase separated at temperatures up to 200 °C. [1]. Thus, the critical molecular weight for domain formation in S-B-S block copolymers appears to be about 7,000. S-iB-S block copolymers behave similarly, but the critical molecular weight is lower, probably about 5,000 [14], because of the greater incompatibility of the polyisobutylene with the polystyrene. These values of the critical molecular weights and temperatures for domain formation agree quite well with the predictions of the theory.

3.1.6 Miscibility with Other Polymers

Because styrenic block copolymers are two-phase systems, when another polymer or an oligomer is added, the new material can be distributed between the phases in several ways. Because almost all end uses of these polymers involve the addition of such polymers or oligomers, this has important practical consequences (see Chapter 7). Materials mix only if the free energy of mixing (ΔG_m in Eq. 3.1) is negative. Thus, the condition for mixing to take place is the reverse of that for domain formation, i.e.,

$$T\Delta S_m > \Delta H_m \tag{3.3}$$

For mixtures of hydrocarbons, ΔH_m is usually positive and ΔS_m varies with the reciprocal of the molecular weight and becomes very small at high molecular weights. For mixtures of hydrocarbon polymers, $T\Delta S_m$ is therefore usually less than ΔH_m. Consequently, most of these polymer pairs are immiscible. Of course, if the two polymers are the same, ΔH_m is zero. In this case, although $T\Delta S_m$ is very small, it is still greater than ΔH_m, and the polymer is miscible with itself.

Block copolymers present a different case. When the effects of adding homopolymers to block copolymers were considered [15], it was shown that even if the homopolymer is structurally identical to one segment of the block copolymer, significant amounts are not be miscible unless the homopolymer molecular weight is much less than that of the corresponding segment in the block copolymer. For example, if a homopolystyrene with a molecular weight of about 100,000 is added to a styrenic block copolymer of similar total molecular weight, the homopolystyrene is not miscible with the polystyrene domains to any significant extent. Instead, it forms a separate phase.

There are restrictions on the size of the polystyrene domains in these block copolymers. A very simple calculation [1] shows that because of these restrictions, the domains are too small to scatter visible light much; thus, although the styrenic block copolymers are two-phase systems, they are transparent. However there are no restrictions on the size of the particles of the added polystyrene and these relatively large particles do scatter light; consequently, mixtures of homopolystyrene and styrenic block copolymers are opaque.

The amount of a homopolystyrene (H) that can dissolve in the polystyrene phase of an S-B block copolymer has been calculated [15] with the results shown in Fig. 3.7. If the molecular weight of the homopolystyrene (M_H) is similar to the molecular weight of the polystyrene segment in the block copolymer (M_S), very little of the added homopolystyrene can mix with the polystyrene domains. Significant mixing between the homopolystyrene and the polystyrene domains occurs only if $M_H << M_S$. The same considerations apply to the elastomer phase. Thus, as a rule, only

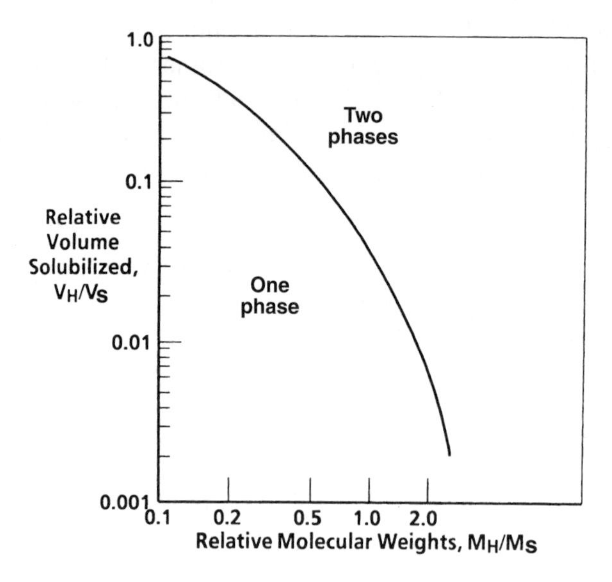

Figure 3.7 Compatibility of an S-B block copolymer with homopolystyrene

very low molecular weight resins and oils should be compatible with either phase of a block copolymer.

An exception is poly(phenylene ether), (PPE), or, as it is sometimes described, poly(phenylene oxide). Because of a positive interaction between the PPE and polystyrene, ΔH_m is negative and so, added PPE of higher molecular weight is miscible with the polystyrene domains in styrenic block copolymers [16]. In other words, we have the apparently paradoxical situation that another polymer (PPE) behaves as though it is more compatible with polystyrene domains than homopolystyrene is with these same polystyrene domains.

The explanation is simple. There is an attraction between PPE and polystyrene, (ΔH_m is negative). There is no attraction between polystyrene and polystyrene (ΔH_m is zero). The miscibility normally found rests on the absence of repulsion. In the case of polystyrene segments mixing with polystyrene, there is a loss in entropy if the added polystyrene goes into the domains. Only if this entropy loss is offset by a gain in entropy from the mixing of small molecules will ΔS_m be positive, allowing mixing to take place.

These effects can be demonstrated very easily. A solution of a styrenic block copolymer in a good solvent is mixed with a small amount of homopolystyrene, e.g., 10% of the total polymer. A drop of this solution is allowed to evaporate on a microscope slide. If the homopolystyrene goes into the polystyrene domains, it will increase their diameter, typically about 300 Angstroms, by about 10%. This increase however, will still leave their size well below the wavelength of light (about 5000 Angstroms) and so the product will be clear. On the other hand, if the homopolystyrene forms a separate phase, there are no restrictions on the size of the particles formed. Homopolystyrene tends to form large particles and so reduce its interfacial surface energy. These large particles will scatter light and so, the product will be opaque. As the theory predicts, the compatibility of the homopolystyrene with the styrenic block copolymer, as judged by the clarity of the mixture, can be improved in two ways. The first is by increasing M_S, i.e., by using a higher molecular weight block copolymer. The second is by decreasing M_H, i.e., by using a lower molecular weight homopolystyrene. Similar effects can be shown using styrenic resins or hydrocarbon oils.

3.1.7 Solubility

Most other types of thermoplastic elastomers have crystalline hard phases and so are insoluble. However, styrenic block copolymers have no

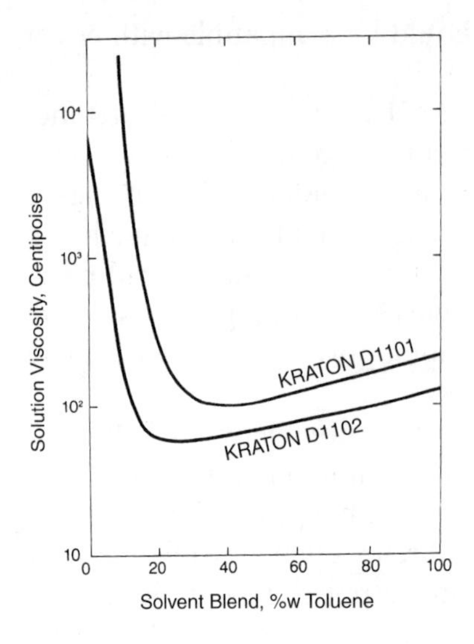

Figure 3.8 Viscosities of 15% solutions of two S-B-S block copolymers in an aliphatic hydrocarbon/toluene mixed solvent

crystallinity and are soluble in most organic solvents. This is both an advantage in that they can be applied from solution, and a disadvantage in that they have poor solvent resistance, although the latter can be improved by compounding. Because styrenic block copolymers are segmented systems, there can be no solvent that is ideal for both segment types. Mixed solvent systems often yield lower solution viscosities than pure solvents, as shown in Fig. 3.8 [17].

3.2 Styrenic Block Copolymers Produced by Anionic Polymerization

These polymers are extensively described in the literature [1–4]. In commercial applications, three elastomeric mid-segments have been produced by anionic polymerization and used for many years: polybutadiene, polyisoprene, and poly(ethylene-butylene). The corresponding block copolymers are referred to as S-B-S, S-I-S, and S-EB-S. Recently, polymers with poly(ethylene-propylene) mid-segments (S-EP-S) have been introduced.

Styrenic block copolymers have the basic structure $(S\text{-}E)x_n$. Here, S represents a polystyrene segment, *E* represents an elastomer segment, and x represents a junction point with a functionality of n. This gives three basic structures: diblock, triblock, and branched ($n=1$, $n=2$ and $n>2$, respectively).

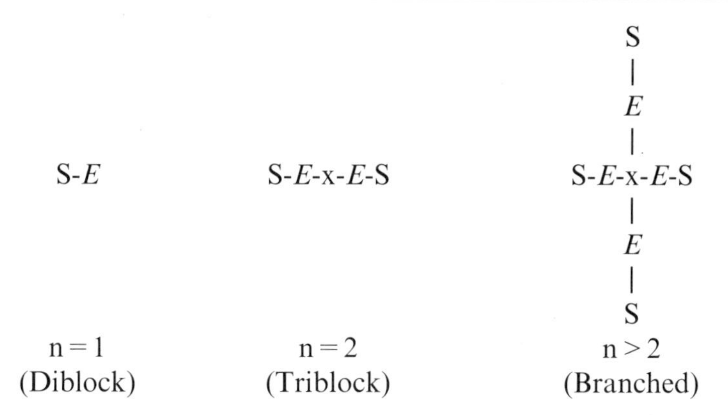

As noted above, diblock polymers are very weak materials because only one end of each polymer chain is anchored in the polystyrene domains. The triblock and branched structures give quite similar properties; both are strong, elastic materials and are generally considered together.

3.2.1 Polymerization

These block copolymers are made by anionic polymerization [18 to 21]. In principle, this is a very simple system in which the polymer segments are produced sequentially from the monomers. It takes advantage of the fact that the system is "living." Thus, a segment of poly(A) can initiate further polymerization of monomers of B, and so on. The first step in the polymerization is the reaction of an alkyl-lithium initiator (R^-Li^+) with styrene monomer.

$$R^-Li^+ + nS \rightarrow R\text{-}S\text{-}S\text{-}S\text{-}S\text{-}S\text{-}S\text{-}S^-Li^+ \qquad (3.4)$$

This product has been termed a "living polymer" because it can initiate further polymerization. If a second monomer, here butadiene, is added:

$$R\text{-}S\text{-}S\text{-}S\text{-}S\text{-}S\text{-}S\text{-}S^-Li^+ + nB \rightarrow R\text{-}S\text{-}S\text{-}S\text{-}S\text{-}S\text{-}S\text{-}S\text{-}B\text{-}B\text{-}B\text{-}B\text{-}B\text{-}B^-Li^+ \qquad (3.5)$$

By repeating these steps, block copolymers with multiple alternating blocks (S-B-S-B-S....) can be produced, but there are no apparent advantages in going beyond triblocks (i.e., S-B-S). After the second polystyrene segment has been completely polymerized, a protonating species (e.g. methyl alcohol) is added to stop the reaction and give the S-B-S block copolymer.

Another variation is to use a coupling reaction to make linear or branched structures such as $(S\text{-}B)_nx$, where x represents an n-functional junction point. A typical example is:

$$2\ (S\text{-}S\text{-}S\text{-}S\text{-}B\text{-}B\text{-}B\text{-}B^-Li^+) + Cl\text{-}R\text{-}Cl \rightarrow S\text{-}S\text{-}S\text{-}S\text{-}B\text{-}B\text{-}B\text{-}B\text{-}R\text{-}B\text{-}B\text{-}B\text{-}B\text{-}S\text{-}S\text{-}S\text{-}S + 2LiCl \tag{3.6}$$

Many coupling agents have been described, including esters, organohalogens, and silicon halides [18, 22]. The example above shows the reaction of a difunctional coupling agent, but those of higher functionality, e.g., $SiCl_4$, can also be used. These give branched or star-shaped molecules such as $(S\text{-}B)_nx$.

The third method of producing these block copolymers uses multifunctional initiation [21, 23, 24]. In this method, a multifunctional initiator ($Li^{+-}R^-Li^+$) is first reacted with the diene (here, butadiene).

$$nB + Li^{+-}R^-Li^+ \rightarrow Li^{+-}B-R-B^-Li^+ \tag{3.7}$$

The final two steps are similar to the corresponding steps in the sequential polymerization described above. When the reaction to produce the $Li^{+-}B-R-B^-Li^+$ is completed, styrene monomer is added and the $Li^{+-}B-R-B^-Li^+$ in turn initiates its polymerization onto the "living" chain ends to give $Li^{+-}S-B-R-B-S^-Li^+$. A protonating species is then added to end the reaction and give the S-B-R-B-S polymer. This example shows the use of a difunctional initiator. There is no reason in principle why initiators of higher functionality could not be used, but none appears to have been reported in the literature.

All these reactions take place only in the absence of terminating agents such as water, oxygen, or CO_2; thus, they are usually carried out under nitrogen and in an inert hydrocarbon solvent. These conditions produce polymers with narrow molecular weight distributions and precise molecular weights.

Only three common monomers – styrene, butadiene, and isoprene – are easy to polymerize using this process, so only S-B-S and S-I-S block copolymers are directly produced on a commercial scale. In both cases, the elastomer segments contain one double bond per molecule of original monomer.

$$\left[CH_2-CH=CH-CH_2-\right]_n \qquad \left[CH_2-\underset{\displaystyle CH_3}{\underset{|}{C}}=CH-CH_2-\right]_n$$

Polybutadiene Polyisoprene

These bonds are quite reactive and limit the stability of the product. More stable analogues can be produced from S-B-S polymers in which the polybutadiene mid-segment consists of both the 1,4 and 1,2 isomers. After hydrogenation, this yields a saturated elastomer equivalent to a copolymer of ethylene and butylene (EB).

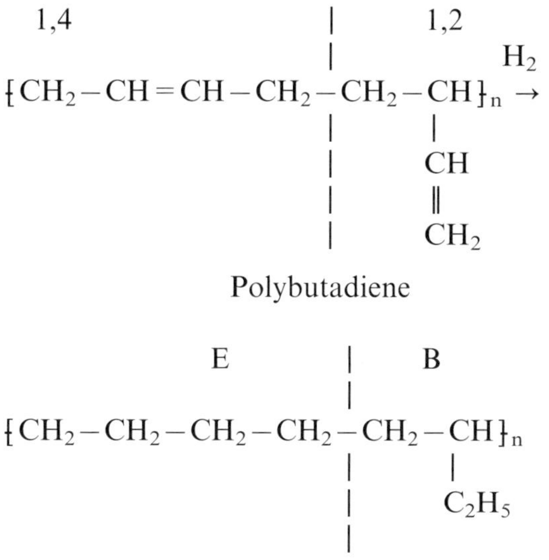

$$\underbrace{\{CH_2-CH=CH-CH_2}_{1,4}-\underbrace{CH_2-\underset{\underset{\overset{\|}{CH_2}}{CH}}{\overset{|}{CH}}\}_n}_{1,2} \xrightarrow{H_2}$$

Polybutadiene

$$\underbrace{\{CH_2-CH_2-CH_2-CH_2}_{E}-\underbrace{CH_2-\underset{C_2H_5}{\overset{|}{CH}}\}_n}_{B} \qquad (3.8)$$

Poly(ethylene-butylene)

Similarly, S-EP-S block copolymers can be produced by hydrogenating S-I-S precursors.

3.2.2 Elastomer Segment

The elastomer segment in the three most important, anionically polymerized, styrenic block copolymers (S-B-S, S-I-S and S-EB-S) have the structures shown above.

Polymers containing poly(ethylene-propylene) elastomer (EP) are newer developments. Their structures are similar to that of poly(ethylene-butylene), with the pendant ethyl group replaced by a methyl group. Their properties are quite similar to those of S-EB-S analogs.

The four block copolymers thus produced (S-B-S, S-I-S, S-EB-S and S-EP-S) show very significant property differences that depend on the nature of the elastomeric mid-segment, as shown in Table 3.2. This also includes data on polyisobutylene (see Section 3.3). These differences affect their suitability for various enduses. The S-B-S polymers are generally used in less demanding applications, such as footwear. The S-I-S polymers are softer and stickier and are often used to produce adhesives. The S-EB-S and

Table 3.2 Comparison of S-B-S, S-I-S, S-EB-S, S-EP-S and S-iB-S Block Copolymers

	Relative Stiffness	Relative Cost	Stability	Degradation Product
S-B-S	1	1	Moderate	Crosslinking
S-I-S	0.5	1.3	Moderate	Chain Scission
S-EB-S & S-EP-S	2	2–2.5	Excellent	Chain Scission
S-iB-S	0.3	1.5?[a]	Excellent	Chain Scission

[a]Estimated S-iB-S polymers are not produced commercially

S-EP-S polymers are the most stable and are used in demanding applications, such as automotive parts and wire insulation.

In early work [1,2], the stiffness and swelling of these polymers was related to the molecular weight between chain entanglements: the stiffer block copolymers are those with the most highly entangled chains. This has been confirmed [25]. A more detailed comparison of the molecular weight between entanglements for these elastomers is given in Table 3.3, which again includes data on polyisobutylene, the least entangled of all (again, see Section 3.3). Consequently, the stress-strain curve of an S-iB-S copolymer [28] lies well below that of a similar S-B-S [12], which in turn lies well below a similar S-EB-S (Fig. 3.9).

Table 3.3 Molecular Weights Between Entanglements (M_e)

Elastomer	M_e (Dynamic)	M_e (Viscous)	M_e (Calculated[c])
Poly(ethylene-butylene) & Poly(ethylene-propylene)	1,660[a]	–	1,530
Polybutadiene	1,700	5,600	2,150
Polyisoprene	6,100[b]	14,000	6,600
Polyisobutylene	8,900	15,200	9,610

[a]Taken from the value quoted for poly(ethylene-propylene)
[b]Taken from the value quoted for natural rubber
[c]Calculated from the plateau moduli of S-*E*-S block copolymers
Results in first two columns taken from Tables 13-I and 13-II of Reference 26. Results in third column taken from Table V of Reference 27. Note that the values obtained from measurements of the viscosity-molecular weight relationship are significantly larger than those obtained dynamically or calculated, but still in the same order.

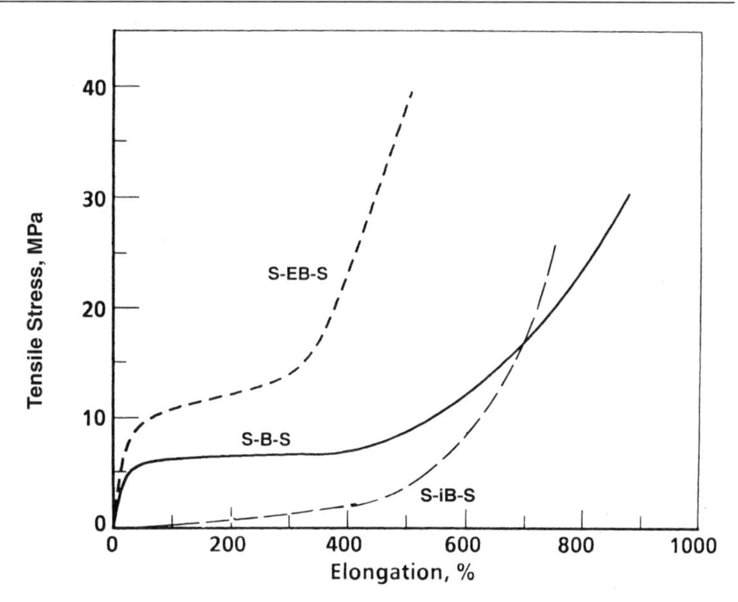

Figure 3.9 Stress–strain properties of styrenic block copolymers having different elastomeric mid-blocks

So far, this section has covered anionically polymerized, styrenic block copolymers with hydrocarbons as the elastomeric segments. These are by far the most important commercially. Other elastomers investigated include poly(propylene sulfide) [4] and poly(dimethyl siloxane) or as it is commonly called, silicone rubber [4, 6]. In both cases, the tensile properties were inferior to those of analogous block copolymers having hydrocarbons as the elastomeric segments [4].

3.3 Styrenic Block Copolymers Produced by Carbocationic Polymerization

A new development that has much commercial potential has recently been described in detail in the literature [5, 29]. These block copolymers are produced by carbocationic polymerization and have several important differences from the anionically polymerized materials described above:

1. The polymerization system
2. The elastomer block can only be polyisobutylene
3. Besides polystyrene, substituted polystyrenes or other polyaromatics can also be used as the end blocks

3.3.1 Polymerization

Carbocationic polymerization is more complex than the anionic system described earlier. The initiators have two or more functionalities and the general formula, $(F\text{-}R)_n x$, where F-R represents a hydrocarbon moiety with a functional group F, and x represents an n-functional junction point. F can be a chlorine, hydroxyl, or methoxy group. Polymerization is carried out at low temperatures of about -80 °C in a moderately polar solvent in the presence of a coinitiator such as $TiCl_4$ or BCl_3. As in anionic polymerization, the polymer segments are produced sequentially from the monomers. Thus, an S-iB-S block copolymer is produced in two stages:

$$F\text{-}R\text{-}F + 2n(iB) \rightarrow {}^{+}(iB)_n\text{-}R\text{-}(iB)_n{}^{+} \tag{3.9}$$

The product, which we can write as ${}^{+}(iB)_n{}^{+}$, is a difunctional living polymer. It can initiate further polymerization, so if a second monomer, such as styrene, is added:

$${}^{+}(iB)_n{}^{+} + 2m(S) \rightarrow {}^{+}(S)_m(iB)_n(S)_m{}^{+} \tag{3.10}$$

It might appear that the very low reaction temperature would present a problem for commercialization. However, a similarly low temperature has been used for many years in the commercial production of polyisobutylene and its copolymers, so there is plenty of practical experience with similar systems.

3.3.2 Elastomer Block

For block copolymers produced by carbocationic polymerization, the only choice for the elastomer block is polyisobutylene, a very unusual material. It is so sterically hindered that it cannot be assembled using standard atomic models. This sterically hindrance results in a relatively inflexible polymer chain. Despite this, polyisobutylene is a rubber, but has a very high molecular weight between chain entanglements (see Table 3.3). Thus, S-iB-S block copolymers, are softer than comparable block copolymers made with polyisoprene, polybutadiene or poly(ethylene-butylene) mid-segments (see Fig. 3.9) and so, are the softest of the styrenic block copolymers. This property should make them very suitable for adhesive applications.

Additionally, polyisobutylene has very low resilience [30], which in turn gives it very high mechanical damping and thus, the ability to absorb vibrations. This characteristic is shared by the S-iB-S block copolymers [5]. Polyisobutylene contains neither double bonds nor tertiary carbon atoms and so the thermal stabilities of S-iB-S and S*-iB-S* block copolymers should be at least equivalent to those of S-EB-S and S-EP-S analogs. The glass transition temperature of polyisobutylene is about –70 °C [5, 31], which is surprisingly low for a material with such very high damping characteristics. This low glass transition temperature means that S-iB-S block copolymers and their compounds should have good low temperature properties. The gas permeability of polyisobutylene is also very low for a rubber [30] and so, S-iB-S block copolymers have good gas barrier properties [5]. The strength properties of S-iB-S block copolymers are similar to the equivalent anionically polymerized block copolymers described in Section 3.1 [5].

3.3.3 Styrenic End Block

Besides as polystyrene itself, a wide range of substituted polystyrenes and other polyaromatics can be used for the end blocks. This is in contrast to the anionically polymerized, styrenic block copolymers, where polystyrene is the only practical choice for the end block [2]. Among the monomers that have been investigated are α-methyl styrene, para-methyl styrene, para-tertiary butyl styrene, para-chloro styrene, indene and acenapthylene, and also copolymers of indene with some of these monomers (see Table 3.4).

Table 3.4 Possible Styrenic and Aromatic End Blocks for S-iB-S and S*-iB-S* Block Copolymers (Reference 5)

End Block Monomer	Polymer Glass Transition Temperature, Tg (°C)	Notes
Styrene	100	Most thoroughly investigated
α-methyl styrene	173	Low cost and high Tg but difficult to polymerize
Para-methyl styrene	108	–
Para t-Butyl styrene	142	Low cost monomer
Indene	200	Low cost monomer, high Tg
Acenapthylene	250	Very high Tg but difficult to polymerize
Para-chloro styrene	129	Should be flame resistant
Para-fluoro styrene	109	–
Indene copolymers[a]	Variable	Can optimize Tg for end-use

[a]With para-methyl styrene and para t-butyl styrene

One potential advantage is that all these substituted polystyrenes have higher glass transition temperatures than polystyrene itself. Thus, S*-iB-S* block copolymers, where S* represents a substituted polystyrene or polyaromatic segment (see Section 3.2.3), should have higher upper service temperatures than either S-iB-S or S-EB-S equivalents.

These unusual and advantageous properties are combined with the fact that S-iB-S and S*-iB-S* block copolymers can be produced directly from readily available, low-cost monomers. This avoids the expensive and inconvenient hydrogenation step described in Section 3.2.1. All these facts should make S-iB-S and S*-iB-S* very competitive with S-EB-S and S-EP-S analogs for many applications. Just why this potential has not been commercially exploited remains somewhat of a mystery.

References

1 Holden, G, Bishop, E. T. and Legge, N. R. (1969) *J. Poly Sci, C26*, 37
2 Holden, G. and Legge, N. R. in Chapter 3, *Thermoplastic Elastomers*, 2nd Ed., G. Holden, N.R. Legge, R.P. Quirk and H.E. Schroeder Eds., (1996), Hanser Publishers, Munich, Vienna, New York
3 Halper, W. M. and Holden, G., in Chapter. 2, *Handbook of Thermoplastic Elastomers*, 2nd Ed., B.M. Walker and C.P. Rader, Eds., (1998), Van Nostrand Reinhold, New York
4 Quirk, R. P. and Morton, M. in Chapter 4 of Reference 2
5 Kennedy, J. P. in Chapter 13 of Reference 2
6 Saam, J. C., Howard, A. and Fearon, F. W. G. (1973) *J. Inst. Rubber Ind. 7*, 69
7 Bi, L. K. and Fetters, L. J. (1975) *Macromolecules 8*, 98
8 Molau, G. E. in *Block Polymers*, S. L. Aggarwal, Ed., (1970) Plenum Press, New York, NY. p. 79
9 Hendus, H., Illers, K. H. and Ropte, E. (1967) *Kolloid Z.Z. Polymere 216–217*, 110
10 Bates, F. S. and Frederickson, G. H. in Chapter 12 of Reference 3
11 Hashimoto, T. in Chapter 15A of Reference 3
12 Gergen, W. P. et al. in Chapter 11 of Reference 3
13a Chung, C. I. and Gale, J. C. (1976) *J. Polym. Sci., Polym. Phys. Ed., 14,* 1149
13b Gouinlock, E. V. and Porter, R. S. (1977) *Poly. Eng. Sci., 17,* 535
14 Gyor, M., Fodor, Z., Wang, H. C. and Faust, R. (1994) *Macrormol. Sci.-Pure Appl. Chem. A31,* 2055
15 Meier, D. J. (1977) *Polym. Prep. 18,* 340
16 Paul, D. R. in Chapter 15C of Reference 2
17 Technical Bulletin SC:72–85 (1985) Shell Chemical Company, Houston
18 Dreyfuss, P, Fetters, L. J. and Hansen, D. R. (1980) *Rubber Chem. Technol. 53*, 738
19 Morton, M., *Anionic Polymerization: Principles and Practice*, (1983) Academic Press, New York, NY
20 Hsieh, H. L. and Quirk, R. P. *Anionic Polymerization: Principles and Practical Applications*, Marcel Dekker, Inc., New York
21 Szwarc, M., Levy, M. and Milkovich, R. (1956) *J. Am Chem. Soc. 78,* 2656

22 Legge, N. R., Davison, S., DeLaMare, H. E., Holden, G. and Martin, M. K. in Chapter 9, *Applied Polymer Science*, 2nd Ed. R. W. Tess and G. W. Poehlein, Eds., (1985) ACS Symposium Series No. 285, American Chemical Society, Washington, D.C.
23a Tung, L. H. and Lo, G. Y-S. (1994) *Macromolecules 27*, 2219
23b Bredeweg, C. J., Gatzke, A. L., Lo, G. Y-S. and Tung, L. H. (1994) *Macromolecules 27*, 2223.
23c Lo, G. Y-S., Otterbacher, E. W., Gatzke, A. L. and Tung, L. H. (1994) *Macromolecules 27*, 2233
23d Lo, G. Y-S., Otterbacher, E. W., Pews, R. G. and Tung, L. H. (1994) *Macromolecules 27*, 2241
23d Gatzke, A. L. and Green, D. P. (1994) *Macromolecules 27*, 2249
24a Tung, L. H., Lo, G. Y-S. and Beyer, D. E. (1980) U.S. Patent 4,196,154 (to Dow Chemical Co.)
24b Tung, L. H., Lo, G. Y-S., Rakshys, J. W. and Beyer, D. E. (1980) U.S. Patent 4,201,729, (to Dow Chemical Co.)
25 Bard, J. K. and Chung, C., I. in Chapter 16 *Thermoplastic Elastomers – A Comprehensive Review*, N.R. Legge, G. Holden and H.E. Schroeder, Eds. (1987) Hanser Publishers, Munich, Vienna, New York
26 Ferry, J. D. *Viscoelastic Properties of Polymers*, 2nd. Ed. (1971) John Wiley & Sons Inc., New York, p.344
27 Tse, M. F., Dias, A. J. and Wang, H.-C. (1999) *Rubber Chem. Technol. 71* 803
28 Faust, R. *Macromol. Chem. Macromol. Symp.* (1994) *85*, 295
29 Matyjaszewski, K. *Cationic Polymerizations: Mechnisms, Synthesis and Applications*, (1996) Marcel Dekker, Inc., New York
30 Fusco, J. V. and Hous, P. in Chapter 10, *Rubber Technology*, 3rd Ed., M. Morton, Ed., (1987) Van Nostrand Reinhold, New York
31 Kresge, E. N., Schatz, R. H. and Wang, H.-C. "Isobutylene Polymers" in *Kirk-Othmer Encyclopedia of Polymer Science and Engineering*, 2nd Ed, J. I. Kroschwitz, Ed. (1986) John Wiley & Sons, Inc., New York

4 Multi-Block Copolymers

These polymers have more complex structures than the styrenic block copolymers. They are based on multi-block $(H\text{-}E)_n$ copolymers, in which the hard (*H*) segments are often crystalline thermoplastics, while the softer (*E*) segments are amorphous and elastomeric.

It is a general principle that crystalline polymers (or segments) must have a regular repeating structure along the polymer chain. Polymers that do not have this regular repeating structure cannot crystallize and so, are amorphous. Thus, the two segment types that form the crystalline multi-block $(H\text{-}E)_n$ polymers have different structures:

- The hard *H* segments are formed from polymers with a *regular repeating* structure,

  -A-A-A-A-A-A-A-A-A-A-A-A-A-A-A-A-

  and with crystal melting points well above room temperature. A high crystal melting point is desirable, because it results in a high upper service temperature in the final product. However if it is too high, the polymer may decompose during melt processing.
- The soft *E* segments are formed from polymers with an *irregular random* structure,

  -A-B-B-A-A-B-A-B-B-B-A-A-B-A-B-B-A-

  These polymers must have glass transition temperatures below room temperature, otherwise they would not be soft and elastomeric. A low glass transition temperature is desirable, because it reduces the lower service temperature of the final product. The difference between the A and B units is sometimes obvious (e.g., A and B are derived from different monomers, such as ethylene and 1-hexene). In other cases, the difference is a little more subtle (e.g., A and B are derived from the same monomer but are arranged in different spatial configurations).

The resultant multi-block $(H\text{-}E)_n$ polymers have a morphology similar to that shown in Fig. 4.1. Here, the hard crystalline segments form an

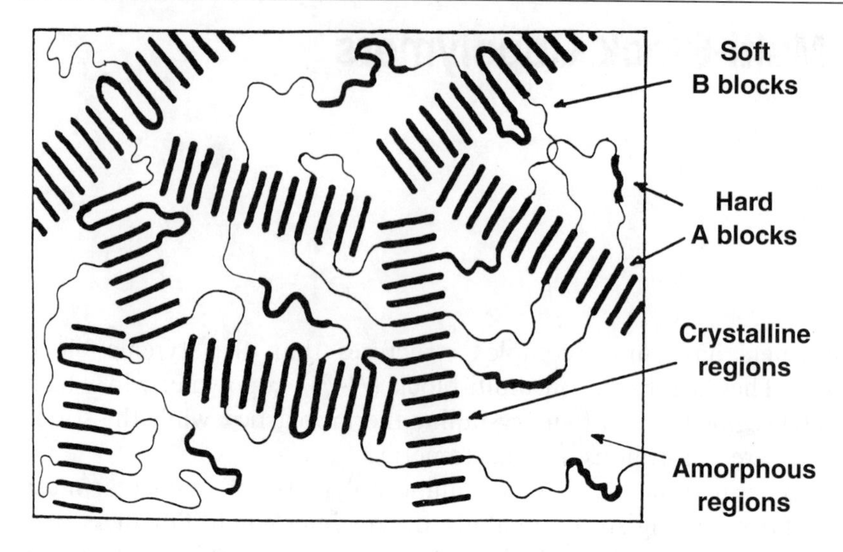

Figure 4.1 Morphology of multiblock polymers with crystalline hard segments

interconnected structure interspersed with the elastomer phase. A single molecule can traverse several crystalline regions. Depending on the perfection of the crystallization process, some hard segment material may be in the elastomer phase and may not crystallize. However, most of the hard segment material forms in separate regions. At room temperature, these crystallize and act as physical crosslinks, tying the elastomeric mid-segments together in a three-dimensional network. In some ways, this is similar to the network formed by the styrenic block copolymers. In both cases, the hard regions lose their strength when the material is heated, allowing the polymer to flow. When the material cools, these regions become hard again and the network regains its original integrity.

In these multiblock $(H\text{-}E)_n$ polymers, the number of segments in the polymer chain and the the distribution of their molecular weights are not as well controlled as those of the styrenic thermoplastic elastomers. Thus, a particular polymer molecule might have five, eight, or twelve segments, all of different molecular weights.

Many these materials have been investigated and are listed, together with other elastomeric block copolymers in Table 4.1. The most important commercial ones can be classified into two groups:

- Polyurethanes, Polyesters, and Polyamides (the descriptions refer to the compositions of the hard segments)
- Polyolefins

Table 4.1 Thermoplastic Elastomers based on Other Block Copolymers

Hard Segment, *H*	Soft or Elastomeric Segment, *E*	Structure[a]	Refs
Polyurethane	Polyester & Polyether	M	[1–3]
Polyester	Polyether	M	[4, 5]
Polyamide	Polyester & Polyether	M	[6, 7]
Polyethylene	Poly(α-olefins)	M	[8, 9]
Polyethylene	Poly(ethylene-co-butylene) and Poly(ethylene-co-propylene)	T	[8, 10]
Polypropylene(isotactic)	Poly(α-olefins)	X	[8]
Poly(β-hydroxyalkanoates)	Poly(β-hydroxyalkanoates)	M	[11]
Polycarbonate	Polydimethylsiloxane	M	[12–14]
Polycarbonate	Polyether	M	[15, 16]
Polyetherimide	Polydimethylsiloxane	M	[17]
Poly(ε-caprolactam), Nylon-6	Polydimethylsiloxane	T	[18]
Poly(methyl methacrylate)	Poly(alkyl acrylates)	T, B	[19]
Polysulfone	Polydimethylsiloxane	M	[20]
Poly(silphenylene siloxane)	Polydimethylsiloxane	M	[21]
Poly(urethane diacetylene)	Poly(ethers)	M	[22]
Polypropylene(isotactic)	Polypropylene(atactic)	X	[8, 9]

a. T = Triblock, *H-E-H*, B = Branched, $(H\text{-}E)_n$x, M = Multiblock, *H-E-H-E-H-*.........
X = Mixed Structures, including multiblock

The polymers in the first group have been produced and sold for many years; in fact, the polyurethanes were one of the first thermoplastic elastomers to be commercialized. Their development took place between 1955 and 1960 [23, 24]. This preceded the discovery of styrenic block copolymers by several years [25]. The similarities in the structures of thermoplastic polyurethane elastomers and styrenic block copolymers (i. e., the existence of hard and soft phases derived from the block copolymer segments) were pointed out in early studies [1]. Thermoplastic elastomers based on polyolefins are a later development.

4.1 Thermoplastic Elastomers Based on Polyurethanes, Polyesters, and Polyamides

In these polymers, the hard crystalline segments *H* are thermoplastic polyurethanes, thermoplastic polyesters, or thermoplastic polyamides. The soft, amorphous *E* segments are usually either polyesters or polyethers, although polycaprolacones, polycarbonates, and mixed copolymers can also be used.

4.1.1 Polymerization

These thermoplastic elastomers are produced from prepolymers by condensation reactions. For those based on polyurethanes and polyesters, three starting materials are used:

1. A long chain diol, also called a polyglycol or polyol ($HO\text{-}R_L\text{-}OH$)
2. A short chain diol, also called a chain extender ($HO\text{-}R_S\text{-}OH$)
3. A difunctional linking monomer (X-R*-X) having groups (X) that can react with the hydroxl groups in the diols

There are two choices for the reactive X groups:

- Isocyanates, which react with the hydroxyl groups to give polyurethanes

$$\sim NCO + HO\sim \rightarrow \sim NHCOO\sim \tag{4.1}$$

- Acids, or their esters, which react with the hydroxyl groups to give polyesters

$$\sim COOH + HO\sim \rightarrow \sim COO\sim + H_2O \tag{4.2}$$
$$\sim COOR + HO\sim \rightarrow \sim COO\sim + ROH \tag{4.3}$$

Thus, the linking momomers are diisocyanates or diacids (or their esters). In the first stage of polymerization, excess linking monomer is reacted with the long chain diol, yielding a prepolymer terminated with the reactive X groups.

$$X\text{-}R^*\text{-}X + HO\text{-}R_L\text{-}OH \rightarrow \underset{\text{Prepolymer}}{X\text{-}(R^*\text{-}\underline{U/E}\text{-}R_L\text{-}\underline{U/E})_n\text{-}R^*\text{-}X} \tag{4.4}$$

where U/E (a urethane or an ester link) represents the product of the reaction of the reactive X groups with the hydroxyl groups of the long chain diol.

We can denote the prepolymer as X-Prepoly-X. It further reacts with the short chain diol and the difunctional linking monomer (the diisocyanate or the diacid):

$$X\text{-Prepoly-}X + HO\text{-}R_S\text{-}OH + X\text{-}R^*\text{-}X \rightarrow X\text{-}[\text{Prepoly-}\underline{U/E}\text{-}(R^*\text{-}U/E\text{-}R_S\text{-}\underline{U/E})_n\text{-}]_m\text{-}R^*\text{-}X \tag{4.5}$$

The final product is a alternating block copolymer with two types of segments:

1. Those formed in the first stage, based on the prepolymer. They are alternating copolymers of the *long* chain diols and the diisocyanate or diacid linking monomers.
2. Those formed in the second stage. These are alternating copolymers of the *short* chain diols and the diisocyanate or the diacid linking monomers.

The long chain diols have a broad molecular weight distribution. Thus, the prepolymers formed from them and the diisocyanate or diacid monomers do not have a regular repeating structure, and are amorphous. Typical glass transition temperatures of the long chain diols are about –45 °C to –100 °C [2] and so, at room temperatures these prepolymers are elastomeric. They form the soft elastomeric phase in the final polymer. In contrast, the short chain diols are a single molecular species (e. g., 1,4-butanediol). Thus, the copolymers formed from them and the diisocyanate or diacid monomers do have a regular repeating structure and can crystallize. Typical crystallization temperatures of these segments are above 150 °C [2, 4] and so, at room temperatures they are hard. They form the hard phase in the final polymer.

There are two ways to produce polyamide thermoplastic elastomers [6]. The first is based on the reaction of a carboxylic acid with an isocyanate to give an amide:

$$\sim NCO + HOOC\sim \rightarrow \sim NHOOC\sim + CO_2 \quad (4.6)$$

Three starting materials are used:

1. A long chain diacid ($HOOC\text{-}R_L\text{-}COOH$)
2. A short chain diacid, also called a chain extender ($HOOC\text{-}R_S\text{-}COOH$)
3. A diisocyanate ($OCN\text{-}R^*\text{-}NCO$)

They react together

$$HOOC\text{-}R_L\text{-}COOH + OCN\text{-}R^*\text{-}NCO + HOOC\text{-}R_S\text{-}COOH \rightarrow$$
$$HOOC\text{-}[R_L\text{-}A\text{-}(R^*\text{-}\underline{A}\text{-}R_S\text{-}\underline{A})_n\text{-}]_m\text{-}R_L\text{-}COOH \quad (4.7)$$

where $\underline{A}$ represents an amide link.

This is similar to the reaction scheme shown above for the production of thermoplastic polyurethane and polyester elastomers. Again, the product is a copolymer with alternating segments. Those derived from the long chain diacids have a broad molecular weight distribution and so, are amorphous. These form the soft elastomeric phase in the final polymer. Those derived from the reaction of the short chain diacids with the diisocyanates have a regular repeating structure and so, are crystalline. These form the hard phase in the final polymer.

In the second method of producing polyamide thermoplastic elastomers, a polyamide terminated by carboxylic acid groups, HOOC-PA-COOH (or the corresponding ester, R′OOC-PA-COOR′), is reacted with a long chain diol:

$$\text{HOOC-PA-COOH} + \text{HO-R}_L\text{-OH} \rightarrow \text{HO-(R}_L\text{-}\underline{\text{E}}\text{-PA-}\underline{\text{E}})_n\text{-R}_L\text{-OH} + 2n\text{H}_2\text{0} \quad (4.8)$$

$$\text{R}'\text{OOC-PA-COOR}' + \text{HO-R}_L\text{-OH} \rightarrow \text{HO-(R}_L\text{-}\underline{\text{E}}\text{-PA-}\underline{\text{E}})_n\text{-R}_L\text{-OH} + 2n\text{R}'\text{OH} \quad (4.9)$$

where E represents an ester link. Essentially, this amounts to preparing an alternating block copolymer from two prepolymers, one (the polyamide) crystalline, the other (the long chain diol) amorphous.

Many different monomers can be used to make alternating block copolymers. The choice determines the properties of the final alternating block copolymer. The following section details some of these effects.

4.1.2 Monomers and their Effects on Polymer Properties

4.1.2.1 Polyurethane Thermoplastic Elastomers

There are many choices for all the three components, the long chain diol, the short chain diol, and the diisocyanate, used to make thermoplastic polyurethane elastomers [16]. Typical examples of the long chain diols are:

- Poly(ethylene adipate) glycol
- Poly(butylene adipate) glycol
- Poly(oxytetramethylene) glycol
- Poly(oxypropylene) glycol
- Polycaprolactone glycol
- Poly(1,6 hexanediol carbonate) glycol

The first two examples are polyesters and are most commonly used. They have a relatively low cost and yield products with good oil resistance. However, the products also have only moderate hydrolytic stability and low temperature flexibility. The next two are polyethers. These are also relatively low cost and yield products with good hydrolytic stability and better low temperature flexibility. However, their oil resistance is not as good as that of the corresponding polyesters. The last two (the poly-caprolactones and the polycarbonates) give premium products with an excellent combination of properties but at a higher price. Typical examples of the short chain diols are

- Ethylene glycol
- Butylene glycol or 1,4-butanediol
- 1,6-hexanediol
- Hydroquinone bis (2-hydroxy ethyl) ether

1,4-butanediol is probably the most important commercially. Replacing it with a diol based on hydroquinone gives products with better upper temperature properties and less compression set. Typical examples of the diisocyanates are

- Diphenylmethane 4,4′-diisocyanate (MDI)
- 2,4 Toluene-diisocyanate (TDI)
- Hexamethylene diisocyanate (HDI)

This list probably represents the order of importance of these materials in commercial products. When combined with a specific short chain diol, MDI gives products with higher crystal melting temperatures. TDI is listed as a suspected carcinogen [26] and this may restrict its use.

4.1.2.2 Polyester Thermoplastic Elastomers

The polyester thermoplastic elastomers [4, 5] are made in a similar way, except that a diacid or diester replaces the diisocyanate. Although many different combinations have been investigated [4], the only one to attain commercial importance is based on a soft segment produced by reacting terephthalic acid (or its methyl diester) with poly- (oxytetramethylene) glycol). The hard segment is produced by reacting terephthalic acid (or its methyl diester) with 1,4-butanediol; in other words, the hard segment is poly(butylene terephthalate). Because the soft segments are ethers and the hard segments are esters, these thermoplastic elastomers are often called polyether esters.

4.1.2.3 Polyamide Thermoplastic Elastomers

The final members of this group are the polyamide thermoplastic elastomers [6, 7]. As described above, these are produced in two ways, either by the reaction of three components (a long chain diacid, a short chain diacid, and a diisocyanate) or by coupling two difunctional prepolymers (a polyamide and a long chain diol). The second method is more important commercially.

In the first case, typical examples of the long chain diacids are:

- Poly(hexamethylene adipate) dicarboxylic acid
- Poly(tetramethylene azelate) dicarboxylic acid
- Poly(oxyethylene) dicarboxylic acid
- Poly(oxypropylene) dicarboxylic acid
- Poly(oxytetramethylene) dicarboxylic acid
- Poly(1,6 hexanediol carbonate) dicarboxylic acid

Polyamide thermoplastic elastomers made from the long chain diacids similar to the first two listed are classified as polyesteramides (PEAs). Those made from diacids similar to the next three listed are classified as polyetheresteramides (PEEAs). Finally, those made from the final type of diacid are classified as polycarbonate-esteramides (PCEAs).

The three types of long chain diacids result in somewhat different properties in the final products. These differences are similar to those produced by changing the type of long chain diols in the polyurethane thermoplastic elastomers. Thus, the polyamide thermoplastic elastomers based on polyesters are relatively low cost and have good oil and oxidation resistance. Those based on polyethers have good hydrolytic stability and low temperature flexibility. Finally, those based on polycarbonates are premium products.

Only two short chain diacids are normally used, adipic (C-6) or azelaic (C-9). The choice is based both on economics and on the melting point of the resulting polyamide segment. This melting point must be high enough so that the hard segments are strong and crystalline at high temperatures but not so high as to increase the processing temperature of the thermoplastic elastomer to the point where it decomposes. Typical melting temperatures are 325 °C for the adipamide and 285 °C for the azelamide. Mixtures of the two diacids can also be used to adjust the melting temperature.

Diphenylmethane 4,4′-diisocyanate (MDI) is the only diisocyanate used to make these polyamide thermoplastic elastomers. Thus, the hard polyamide sequences in PEAs, PEEAs, and PCEAs are aromatic in nature.

Polyester-block-amides (PE-b-As) are polyamide thermoplastic elastomers produced by coupling two difunctional prepolymers, a polyamide terminated with carboxylic acid groups and a long chain diol.

The polyamides used to produce PE-b-As are aliphatic. Some examples are:

- Nylon-6
- Nylon-11
- Nylon-12
- Nylon-6,6
- Nylon-6,12

These have melting points ranging from 180 to 255 °C and differ in the extent of their crystallinity. Thus, products can vary in such properties as upper service temperature and hardness.

The long chain diols used to make PE-b-As are always polyethers; polyesters would undergo an ester-amide interchange with the aliphatic polyamide hard segments. Typical examples of these long chain diols are:

- Poly(oxyethylene) glycol, PEO
- Poly(oxypropylene) glycol, PPO
- Poly(oxytetramethylene) glycol, PTMO

Those based on poly(oxyethylene) glycol have the lowest glass transition temperature but absorb more water. Those based on poly(oxytetramethyllene) glycol have a higher glass transition temperature but absorb less water.

4.1.3 Common Features

The polyurethane-, polyester-, and polyamide-based thermoplastic elastomers share several features. All are hard (usually greater than 70 Shore A), and the hardness can be adjusted by varying the ratio of hard to soft segments in the polymer. They are tough materials with good resistance to cut growth and flex cracking. Prices are fairly high (see Table 8.2) with the polyurethanes the least expensive and the polyamides the most. All have high upper service temperatures and good resistance to oils and greases. When molded or extruded by the fast processes used commercially, crystallization of the hard sements is often incomplete. Thus, properties can often be improved by annealing.

4.2 Thermoplastic Elastomers Based on Polyolefins

These thermoplastic elastomers are also $(H\text{-}E)_n$ block copolymers in which the hard H segments are crystalline and the soft E segments are amorphous.

Both hard and soft segments are polyolefins. The differences in their properties derive from the structural differences in the polymers that make up the segments. These polymers are based either on α-olefins (substituted ethylenes, $CH_2{=}CHR$, where R represents an alkyl group), or on ethylene itself.

4.2.1 Segmental Structures

Formation of hard crystalline segments in these polymers requires the presence of regular repeating units along the polymer chain. For poly (α-olefins), there are two simple, regular stearic arrangements that can crystallize: isotactic (much more common) and syndiotactic:

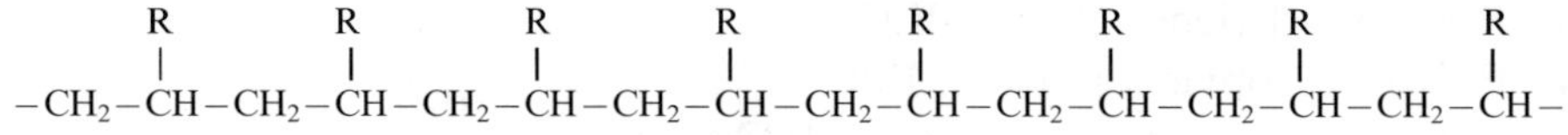

$$-CH_2-\underset{}{\overset{\overset{\displaystyle R}{|}}{CH}}-CH_2-\overset{\overset{\displaystyle R}{|}}{CH}-CH_2-\overset{\overset{\displaystyle R}{|}}{CH}-CH_2-\overset{\overset{\displaystyle R}{|}}{CH}-CH_2-\overset{\overset{\displaystyle R}{|}}{CH}-CH_2-\overset{\overset{\displaystyle R}{|}}{CH}-CH_2-\overset{\overset{\displaystyle R}{|}}{CH}-CH_2-\overset{\overset{\displaystyle R}{|}}{CH}-$$

Isotactic

$$CH_2-\overset{\overset{\displaystyle R}{|}}{CH}-CH_2-\underset{\underset{\displaystyle R}{|}}{CH}-CH_2-\overset{\overset{\displaystyle R}{|}}{CH}-CH_2-\underset{\underset{\displaystyle R}{|}}{CH}-CH_2-\overset{\overset{\displaystyle R}{|}}{CH}-CH_2-\underset{\underset{\displaystyle R}{|}}{CH}-CH_2-\overset{\overset{\displaystyle R}{|}}{CH}-CH_2-\underset{\underset{\displaystyle R}{|}}{CH}-$$

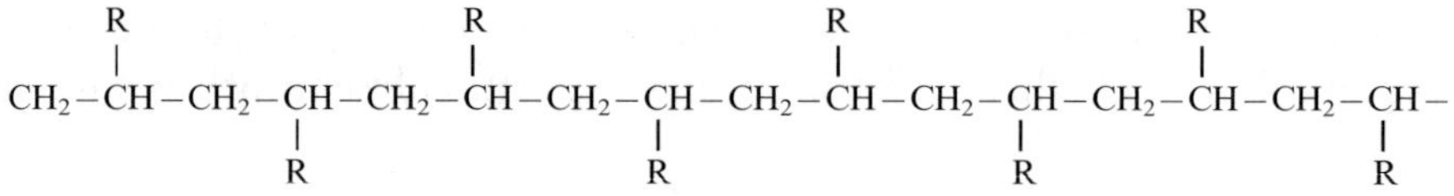

Syndiotactic

For polyethylene, where the pendant group R is replaced by a hydrogen atom, there is only one possible form:

$$-CH_2-CH_2-CH_2-CH_2-CH_2-CH_2-CH_2-CH_2-CH_2-CH_2-CH_2-CH_2-CH_2-CH_2-$$

Polyethylene

The arrangement shows the structure of a perfectly linear polyethylene. Commercially produced high density polyethylene (HDPE) is very close to this structure, although there are always a few side branches. HDPE has a high degree of crystallinity, typically about 70%, and a crystal melting point of about 135 °C. The 30% of the material not crystallized in HDPE has an amorphous structure, with a glass transition temperature that is a matter of some dispute, but is often estimated as about –100 °C. If small amounts of α-olefins are randomly copolymerized into the chain, this reduces both the extent of crystallinity and also the melting point of the crystalline region.

All three of the structures described above have regular repeating units and so can form the hard *H* segments in the polymer chain. On the other hand, if the polymer chain does not have a regular repeating structure, it cannot crystallize and so, is amorphous. At a low enough temperature it goes through a glass transition and becomes hard. However, the glass transition temperatures of poly(α-olefins) are below room temperature. Thus, at room temperature, amorphous poly(α-olefins) are soft and so, can form the elastomeric *E* segments in the block copolymer chain. There are two common amorphous structures in poly(α-olefins): atactic homopolymers and random copolymers:

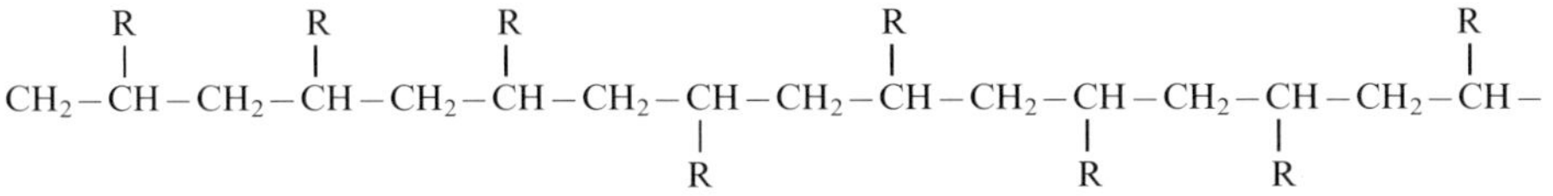

$$CH_2-\overset{R}{\overset{|}{CH}}-CH_2-\overset{R}{\overset{|}{CH}}-CH_2-\overset{R}{\overset{|}{CH}}-CH_2-\underset{R}{\underset{|}{CH}}-CH_2-\overset{R}{\overset{|}{CH}}-CH_2-\underset{R}{\underset{|}{CH}}-CH_2-\underset{R}{\underset{|}{CH}}-CH_2-\overset{R}{\overset{|}{CH}}-$$

Atactic Homopolymer

$$-CH_2-\overset{R}{\overset{|}{CH}}-CH_2-CH_2-CH_2-\overset{R}{\overset{|}{CH}}-CH_2-\underset{R}{\underset{|}{CH}}-CH_2-CH_2-CH_2-CH_2-CH_2-\underset{R}{\underset{|}{CH}}-CH_2-\overset{R}{\overset{|}{CH}}-CH_2-CH_2-$$

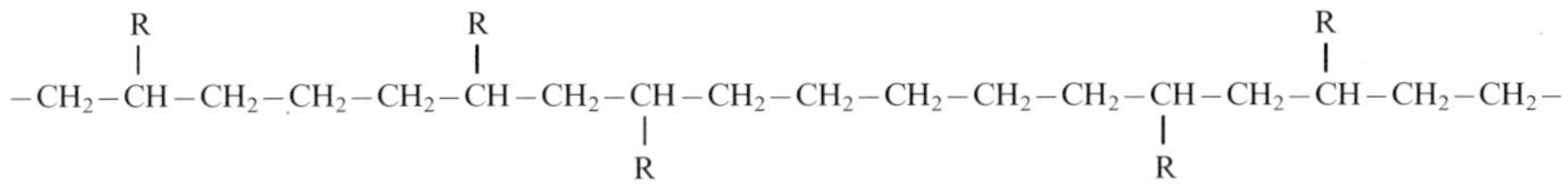

Random Copolymer

This shows a random copolymer of ethylene and an α-olefin. In this case, the function of the α-olefin units is to break up the poly-ethylene sequences so that they cannot crystallize. The best-known example of this type of elastomer is EPR, a random copolymer of ethylene and propylene.

Besides propylene, other important α-olefins monomers are oligamers of ethylene, i.e. molecules that have an even number of carbon atoms (e.g., 1-butene, 1-hexene,1-octene etc). Thus, the choice of polymer for the hard and soft segments for potential thermoplastic elastomers is limited to homopolymers or copolymers of ethylene, propylene, and even numbered α-olefins. Table 4.2 provides the glass transition temperatures and crystal melting points of polyethylene and various poly(α-olefins). The ideal polyolefin-based, thermoplastic elastomer would have hard *H* segments with a high crystal melting point and soft *E* segments with a low glass transition temperature. In practice, this means there are two choices for the hard *H* segments:

- Polypropylene. This polymer, in its isotactic form, has the highest crystal melting point (165 °C) of any of the common poly(α-olefins). This temperature is high enough to result in excellent resistance to heat distortion but it is not so high that it causes significant degradation during melt processing, which must be carried out at some temperature

Table 4.2 Glass Transition and Crystal Melting Temperatures for Poly(α-Olefins)

Polymer	Glass Transition Temperature T_g (°C)	Crystal Melting Temperature T_m (°C)
Polyethylene	– 100[a]	135
Polypropylene	– 10	165*
Poly(1-butene)	– 25	130*
Poly(1-hexene)	– 35	50*
Poly(1-octene)	– 55	20*
Poly(ethylene-co-propylene)	– 60	–

*For isotactic polymers
[a]There is no general agreement on the exact value, but it is low.

significantly above the crystal melting point. Polypropylene can be polymerized in the isotactic form using stereo-specific, Ziegler-Natta catalysts, which has led to its worldwide production and use [27].

- Polyethylene. This is in many ways an ideal choice. The stereo-specific catalysts are not necessary, because pure polyethylene can exist in only one form, linear. As a result, polyethylene has a regular repeating unit that will crystallize.

There are also only two choices for the soft *E* segments:

- Atactic Poly (α-olefins) The disadvantage of many of these materials is that they have fairly high glass transition temperatures (see Table 4.2) and so, relatively poor low temperature properties. The glass transition temperatures drop as the length of the pendant chain increases. Thus, poly (1-hexene) and poly(1-octene) are the best choices; however, their monomers are relatively expensive compared with propylene and 1-butene.
- Ethylene/α-Olefin Copolymers. These have glass transition temperatures intermediate between those of the polymer of the α-olefin and the amorphous phase of polyethylene (about – 100 °C). Thus, although atactic polypropylene has a high glass transition temperature (– 10 °C), the copolymer of propylene with ethylene (EPR) has a much lower one (about – 60 °C).

4.2.2 Polymerization

In principle, the most desirable, general purpose, thermoplastic elastomer would probably be a triblock copolymer or multiblock copolymer of isotactic polypropylene and ethylene/propylene copolymer, i.e., iP-EP-iP or $(iP\text{-}EP)_n$. The crystalline polypropylene segments would give these polymers a high upper service temperature and some degree of solvent resistance. The EP elastomeric block(s) would be flexible at low temperatures and readily extended with hydrocarbon plasticizing oils. The polymers would be easy to process and produced from low-cost monomers. Despite these important potential advantages however, thermoplastic elastomers of this type have not been produced commercially. Although Ziegler–Natta catalysts have been used to produce these and similar copolymers experimentally [8], the results were disappointing. Polymerization was often slow and inefficient and the products were poorly defined mixtures. In contrast, blends and dynamic vulcanizates based on isotactic polypropylene and ethylene/propylene copolymers have been very successful commercially (see Chapter 5).

Block copolymers in which the hard phase is polyethylene have been commercially successful. These are produced using metallocene catalysts, which are typically based on cyclopentadienyl groups linked to a halide of a transition metal, e.g., Ti, Zr, Hf [8, 9, 27]. Under the right conditions, these polymerize mixtures of ethylene and α-olefin monomers into polymers with long repeating ethylene sequences, rather than into the typical random copolymers. The α-olefins are usually 1-butene, 1-hexene, or 1-octene, yielding copolymer segments with pendant groups, usually arranged atactically. Because of their irregular structures, these segments cannot crystallize. Instead, they are amorphous materials with low glass transition temperatures and are soft and rubberlike at room temperature. In contrast, the long polyethylene segments in the polymer chain cannot have irregularities and so can crystallize and form the hard phase.

Based on a value of 68 °C reported for the crystal melting point of the ethylene segment in one such block copolymer [8], the segments are either very short or contain occasional units of the α-olefin comonomer.

Another route in the production of block copolymers based on olefins is the complete hydrogenation of block copolymers of butadiene and isoprene [8, 10]. Polymers such as B-I-B and $B\text{-}B_{50}\text{-}B$ (where B_{50} represents a polybutadiene segment with 50% of the units in the 1,2 configuration) can be produced by anionic polymerization (see Section 3.1.1). These can be hydrogenated to give E-EP-E and E-EB-E, respectively. The melting point of the polyethylene segment is reported to be 107 °C [10], which is significantly higher than the value of 68 °C reported for the polyethylene segment in the ethylene/α-olefin block copolymers described above but less than that of pure linear polyethylene (135 °C). The lower value for the melting temperature of the polyethylene segments in the E-EP-E and E-EB-E block copolymers is caused by the structure of the polybutadiene segments from which they are derived. These contain about 10% of the monomer units in the 1,2 configuration. When hydrogenated, these become 1-butene units and disrupt the crystallinity of the polyethylene, or more correctly, poly (tetramethylene), segments.

The room temperature properties of E-EP-E and E-EB-E block copolymers are reported to be very similar to those of S-B-S and similar styrenic block copolymers [8, 10]; high temperature properties are better. This is probably because the glass transition in polystyrene is a more gradual process (and so starts at a lower temperature) than the melting of the polyethylene crystallites.

Despite these apparently improved properties, E-EP-E and E-EB-E block copolymers have not been commercialized. However, they serve as useful model polymers for other systems.

4.3 Miscellaneous Block Copolymers

As pointed out in a paper dealing with the early work on styrenic thermoplastic elastomers [28], the explanation of their structure/property relationships can be generalized to include all block copolymers with alternating hard and soft segments. Specifically, this explanation was applied to existing polycarbonate/polyether [15, 16] and poly (silphenylenesiloxane)/poly(dimethylsiloxane) [21] block copolymers, both of which are alternating $(H\text{-}E)_n$ block copolymers. (The polymer forming the hard segment is the one listed first). Since that time, many other block copolymers with the general formula *H-E-H* or $(H\text{-}E)_n$ have been shown to be thermoplastic elastomers. Some examples are give in Table 4.1. They include block copolymers with polysiloxane (or silicone rubber, as it is often called) elastomeric segments, which should have excellent thermal stability. Block copolymers with polysiloxane elastomeric segments and poly(etherimide) hard segments have even better thermal stability, although they are very high in price (about $20/lb). These block copolymers are used in critical applications, such as fire resistant wire covering [17]. Even if they degrade during a fire, they do not generate much smoke. Triblock copolymers with polysiloxane elastomeric segments and Nylon-6 hard segments should also have improved thermal stability [18].

Block copolymers based on (β-hydroxyalkanoates) are produced by bacteria [11]. They have the general structure:

$$-[-O-\underset{\substack{| \\ R}}{CH}-CH_2-\underset{\substack{\| \\ O}}{C}-]_n-$$

where R represents a pendant alkyl group. These alkyl groups are arranged isotactically and so the polymers are able to crystallize. However, under the right biosynthesis conditions, pendant alkyl groups of different sizes are incorporated in the polymer chain. These reduce the crystallinity to the point where an amorphous phase is formed also. This gives a two-phase system and so the products are thermoplastic elastomers.

References

1 Cooper, S. L. and Tobolsky, A. V. (1966) *Textile Research Journal, 36,* 800
2 Mekel, W., Goyert, W. and Wieder, W. in Chapter 2, *Thermoplastic Elastomers*, 2nd Ed., G. Holden, N.R. Legge, R.P. Quirk and H.E. Schroeder Eds. (1996), Hanser Publishers, Munich, Vienna, New York
3 Ma, E. C. in Chapter 7, *Handbook of Thermoplastic Elastomers*, 2nd Ed., B.M. Walker, and Rader, C.P.., Eds. (1998), Van Nostrand Reinhold, New York
4 Adams, R. K., Hoeschele, G. K. and Witsiepe, W. K. in Chapter 8 of Reference 2
5 Sheridan, T. W. in Chapter 6 of Reference 3
6 Nelb, R. G. and Chen, A. T. in Chapter 9 of Reference 2
7 Farrissey, W. J. and Shah, T. M. in Chapter 8 of Reference 3
8 Kresge, E. N. in Chapter 5 of Reference 2
9 Laird, J. L. (1997) *Rubber World, 217(1)* 42
10 Quirk, R. P. and Morton, M. in Chapter 4 of Reference 2
11 Gagnon, K. D. in Chapter 15B of Reference 2
12 Vaughn, H. A. (1969) *J. Poly. Sci. B7*, 569
13 Kambour, R. P. (1969) *J. Poly. Sci. B7*, 573
14 LeGrand, D. G. (1969) *J. Poly. Sci. B7*, 579
15 Goldberg, E. P. (1964) *J. Poly. Sci. C4*, 707
16 Perry, L.K., Jackson, W. J. Jr. and Caldwell, J. R. (1965) *J. Appl. Poly. Sci. 9,* 3451
17 Mihalich, J., paper presented at the 2nd International Conference on Thermoplastic Elastomer Markets and Products sponsored by Schotland Business Research, Orlando, FL, March 15–17, 1989
18 Lovinger, A.J., Han, B. J., Padden, F. J. Jr. and Mirau, P. A. (1993) *J. Polym. Sci.: Polym. Phys. 31*, 115
19 Jerome, R. et al. in Chapter 15D of Reference 2
20 Noshay, A., Matzner, M. and Merriam, C. N. (1971) *J. Poly. Sci. Part A1 9*, 3147
21 Merker, R. L., Scott, M. J. and Haberland, G. G. (1964) *J. Poly. Sci. A, 2,* 31
22 Hammond, P. T. and Rubner, M. F. in Chapter 15E of Reference 2
23a Schollenberger, C. S. (1955) U. S. Patent 2,871,218 (to B. F. Goodrich)
23b Schollenberger, C. S., Scott, H. and Moore, G. R. (1958) *Rubber World 137* 549
23c Schollenberger, C. S., Scott, H. and Moore, G. R. (1962) *Rubber Chem. Technol. 35*, 742
24 Charch, W. H. and Shivers, J. C. (1959) *Textile Research Journal 36* 536
25 Holden, G., Legge, N. R., Quirk, R. P. and Schroeder, H. E. in Chapter 1 of Reference 2

26 Section 16, pages 20 and 37, *Handbook of Chemistry and Physics*, 76th. Ed. D. R. Linde, Ed. (1995) CRC Press – Boca Raton/New York/London/Tokyo
27 Kresge, E. N. (1997) Rubber World, *217(1)*, 30
28 Holden, G., Bishop, E. T. and Legge, N. R. (1969) *J. Poly. Sci. C26,* 37

5 Hard Polymer/Elastomer Combinations

There are two types of these materials: simple blends of the two polymers (the hard polymer and the elastomer) and dynamically vulcanized products in which the elastomer is crosslinked during the mixing process. Examples of both types are shown in Table 5.1. Both the hard polymers and the elastomers used to make these products can be obtained "off the shelf." Thus, an almost unlimited range of combinations can be investigated quickly and easily. Similarly, commercial products can be made without the very high capital investment usually required to produce new polymers.

Table 5.1 Thermoplastic Elastomers Based on Hard Polymer/Elastomer Combinations

Hard Polymer *H*	Soft or Elastomeric Polymer *E*	Structure[a]	Refs
Polypropylene	EPR or EPDM	B	[1–4]
Polypropylene	EPDM	DV	]3, 5, 6]
Polypropylene	Poly(propylene/1-hexene)	B	[3]
Polypropylene	Poly(ethylene/vinyl acetate)	B	[3]
Polypropylene	Polyisobutylene	DV	[5, 7]
Polypropylene	Polyisobutylene/Polychloroprene	DV	[5]
Polypropylene	Natural Rubber	DV	[5, 8]
Polypropylene	Nitrile Rubber	DV	[5, 6]
Nylon	Nitrile Rubber, Polyurethane or Epichlohydrin Rubbers	DV	[5]
PVC	Nitrile Rubber + DOPb	B, DV	[9–11]
PVC	Polyurethane or Polyester Thermoplastic Elastomers + DOP[b]	B	[9–11]
Halogenated Polyolefin	Ethylene Interpolymer	B	[11, 12]
Polybutylene terephthalate	EPDM	B, DV	[3]
Polystyrene	S-B-S + Oil	B	[13]
Polypropylene	S-EB-S + Oil	B	[13]

[a]B = Simple Blend, DV = Dynamic Vulcanizate
[b]DOP = Dioctyl phthalate. Other plasticizers could also be used.

5.1 Simple Blends

Most simple blends are produced by mixing the hard polymer and the elastomer together on high shear compounding equipment. An exception is the manufacture of "in-situ" blends of polypropylene and ethylene-propylene rubber in the same reactor. Many possible morphologies can be formed in a two-phase system (see Section 3.1.3 and Fig. 3.4. For the purposes of this chapter, we specify hard polymer/elastomer blends as having one of the following three structures:

1. A dispersion of the elastomer in a continuous phase of the hard polymer. Here the stiffness and strength of the hard polymer predominate. The result is a flexible, tough thermoplastic, typified by high impact polystyrene.
2. This is the reverse of 1, i.e. a dispersion of the hard polymer in a continuous phase of the elastomer. Here the stiffness and strength of the elastomer phase predominate. The result is essentially a filled, unvulcanized elastomer, in which the dispersed particles of the hard polymer act as the filler. Because the elastomer is unvulcanized, it has little strength and the product is soft but too weak for practical applications.
3. A three-dimensional, co-continuous structure of the hard polymer and the elastomer. This structure can be visualized as similar to an open cell foam, in which the foamed material is one phase and air is another. It is represented two dimensionally in Fig. 5.1. This structure yields a product with strength derived from the continuous hard phase, but with flexibility derived from the continuous soft phase. Because neither phase is crosslinked, both can flow; in other words, the result is a thermoplastic elastomer.

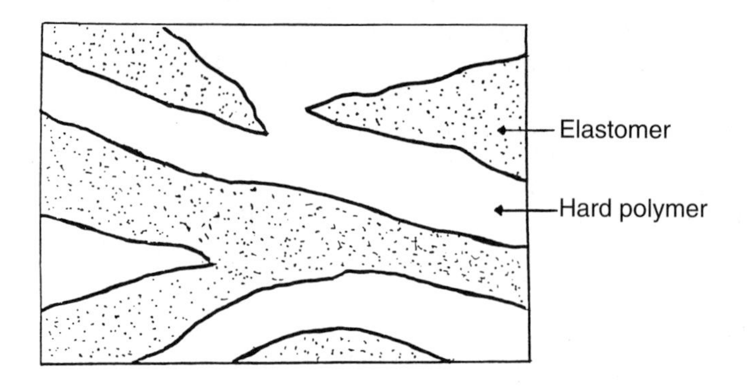

Figure 5.1 Morphology of Hard Polymer/Elastomer Blends

Structure 3 is critical for the formation of a thermoplastic elastomer from a simple two-component blend. There are several requirements for its formation:

1. The viscosities of the two polymers must be matched at the temperature and shear rates of mixing. The temperature during mixing can be controlled by adjusting the heating (and possibly cooling) system(s) on the mixing equipment. The shear rate is in the range of 100 to 1000 sec^{-1}, with the lower end of this range being typical of rubber processing equipment, such as Banbury internal mixers and rubber mills. Conversely, the upper end of this range is typical of plastics processing equipment, such as twin screw extruders. The optimum viscosity match also depends on the proportions of the two components. Figure 5.2 [14] shows a relationship between the ratio of the viscosities of two components (A and B) and the proportion of these components in the mixture. This approach predicts that an ideal viscosity match exists when:

$$\log_{10}(\eta_A/\eta_B) = 2\text{–}4\phi_B \qquad (5.1)$$

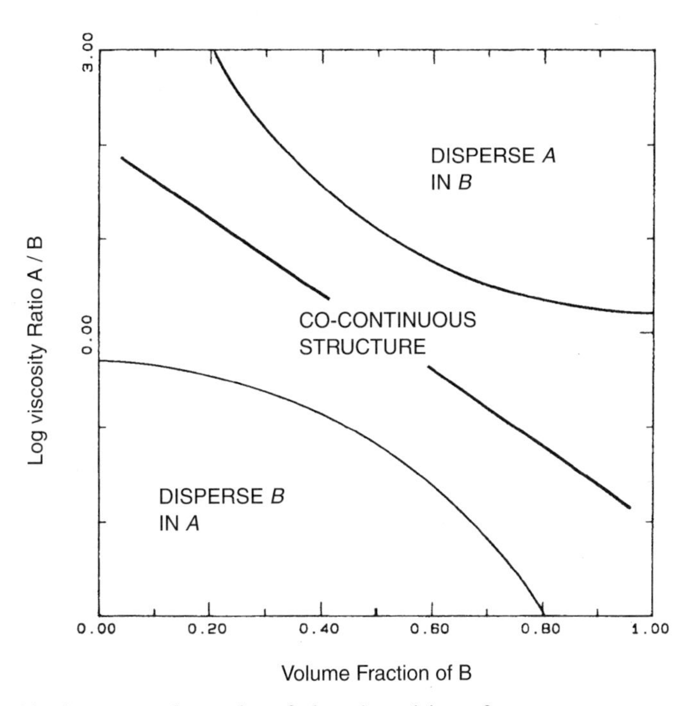

Figure 5.2 Relationship between the ratio of the viscosities of two components (A and B) and their proportion in the mixture

where η_A and η_B are the viscosities of the components A and B at the processing conditions and ϕ_B is the volume fraction of component B. Other workers [15] have used the expression

$$\eta_A/\eta_B = \phi_A/\phi_B \tag{5.2}$$

as the criterion for ideal mixing.
In practice, these two expressions give similar results unless the volumes of the two components are very different. Both predict that if the volumes of the two components are about equal, then for ideal mixing, their viscosities should also be about equal. If the volumes are not equal, then the component with the larger volume should also have the higher viscosity. For example, if the volume of A is about twice that of B, then the viscosity of A should be from two to five times greater than that of B. Increasing the ratio of the viscosities to well above the optimum range favors the formation of a dispersed phase of the higher viscosity component. Decreasing the ratio of the viscosities to well below the optimum range favors the formation of a dispersed phase of the lower viscosity component.

2. The other important factor is the compatibility of the two components, which is often judged by the difference between their solubility parameters. This difference can be correlated with the interfacial tension between the two polymers. If two polymers (or two immiscible liquids) have a large interfacial tension, the formation of a two-phase system with large phase dimensions (i.e., a coarse dispersion) is favored. This coarse dispersion reduces the interfacial area and hence, the interfacial energy. In contrast, polymer pairs with similar solubility parameters tend to form a finer dispersion, as shown in Fig. 5.3 [14]. This graph shows the particle size of a number of S-EB-S/polymer blends. The particle size correlates well with the solubility parameter difference between the EB (ethylene–butylene copolymer) and the other polymer. Thus, if the two polymers have very different solubility parameters, i.e., one is polar while the other is not, they will probably form a coarse dispersion with poor adhesion between the phases.

Thermoplastic elastomer blends based on polypropylene are very significant commercially. Polypropylene has the advantages of being low in both cost and density. Additionally, its crystalline structure and relatively high crystal melting point (T_m = 165 °C) give it resistance to oil, solvents, and high temperatures. EPR or EPDM rubbers are the obvious choices for the

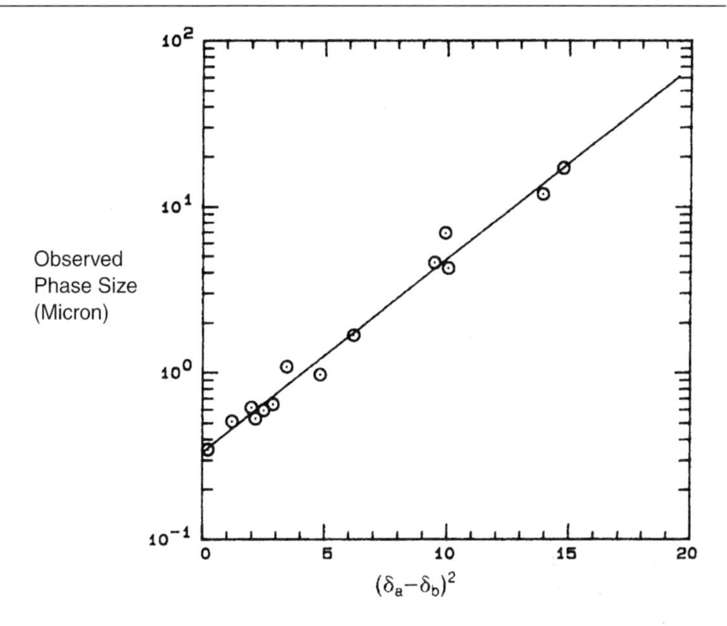

Figure 5.3 Effect of solubility parameter matching on phase size

corresponding elastomer phase because of their thermal stability, low cost, low temperature flexibility, and structural similarity to (and hence, technical compatibility with) polypropylene. Thus, well dispersed blends of polypropylene with EPR or EPDM rubbers are low cost, easy to make, and have a wide service temperature range.

At first, EPDM rubbers were more commonly used, because they were more readily available. They are used in very large amounts as conventionally vulcanized rubbers. However, EPDM rubbers are more expensive than EPR rubbers because of the diene co-monomer in their structure. This extra co-monomer is several times more expensive than either ethylene or propylene (probably the two lowest cost monomers in existence) and is only required if the rubber is to be vulcanized. It is not vulcanized in the simple blends described in this section. Thus, EPR rubbers are a more logical choice blends with polypropylene, if suitable ones are available.

As the applications for these blends have increased, it has become more economical for manufacturers to polymerize the lower cost EPR rubbers specifically for this purpose. In some cases, the polypropylene and the EPR can be produced in the same reactor. In all these blends, hydrocarbon mineral oil can be added to improve processing [1]. Copolymers of propylene with 1-hexene and poly(ethylene-vinyl acetate) have also been investigated as the rubbery phase [3] but have not become commercially important, probably because of the economic advantage of EPR. Typical blends contain 60 to 80% EPDM or EPR (sometimes including hydrocarbon extender oil), with the remainder being polypropylene [1, 3].

Styrenic thermoplastic elastomers are not usually considered as hard polymer/elastomer combinations. However, when they are processed as thermoplastics (e.g., by injection molding, extrusion etc.), they are always blended with hard thermoplastics. Typical combinations are:

- Polystyrene/(S-B-S + oil)
- Polypropylene/(S-EB-S + oil)

Large amounts of filler are often added also. More details of these blends are given in Chapter 7. Note that because the elastomer phases in these combinations are themselves thermoplastic elastomers, they are much stronger than the unvulcanized elastomers used to prepare blends such as polypropylene/EPR. Thus, large amounts of (S-B-S + oil) or (S-EB-S + oil) can be blended with polystyrene or polypropylene to give very soft blends with acceptable properties.

An entirely different class of thermoplastic elastomers is based on blends of poly(vinyl chloride) (PVC) as the hard phase with nitrile rubber (NBR) and a plasticizer, often dioctyl phthalate (DOP). This technology has been known for more than 50 years [16] and is an extension of the even older technology to produce plasticized PVC [17]. These blends are at the ill-defined interface between flexible thermoplastics and thermoplastic elastomers. Although there is no hard and fast dividing line, plasticized PVC is usually considered a flexible thermoplastic rather than a thermoplastic elastomers because of its lack of true elastomeric character. For example, it does not have a "rubbery" feel and frictional characteristics are inferior to those of true rubbers [18]. It does not retract immediately when extended and released, i.e., it has very poor "snapback." The addition of NBR results in more rubberiness and the tri-component blends are much more elastomeric. Special types of NBR have been developed for these blends [9 to 11]. The acrylonitrile content of the NBR, its molecular weight, and its degree of branching can be varied to suit the end use requirements. A typical formulation [11] contained:

- PVC 100 parts
- NBR 33 parts
- DOP 75 parts
- Filler 20 parts

The phase morpologies of these and similar systems appear to be complex. Evidence has been reported for both single-phase and two-phase arrangements [11].

Multi-block polyurethane and polyester thermoplastic elastomers can also be used as the soft phase in plasticized PVC blends [11, 19]. In both cases, the polar character of the thermoplastic elastomer results in a fine dispersion with the PVC and may even yield a single-phase system.

Blends of a halogenated polyolefin with an ethylene interpolymer are reported to give a single-phase system [11, 12]. The products have been commercialized under the Tradename of ALCRYN and are soft, oil resistant, thermoplastic elastomers.

Polybutylene terephthalate (PBT) has many of the advantages of polypropylene. Although it several times the price (on an equal volume basis), it has a higher crystal melting point of about 225 °C [20]. It has been used as the hard phase in both blends and dynamic vulcanizates with EPDM as the elastomer phase [3]. Satisfactory properties were obtained only when about 3% of a graft copolymer of EPDM and glycidal methacrylate was used as a compatibilizing agent. The use of this and similar graft copolymers to improve interphase adhesion is discussed in more detail in the following section.

5.2 Dynamic Vulcanizates

The structure of dynamic vulcanizates [5] is quite different. It consists of a fine dispersion of a crosslinked (or, in the terminology of the rubber industry, vulcanized) elastomer phase in a matrix of a hard thermoplastic (Fig. 5.4). The unvulcanized elastomer, the vulcanizing ingredients, and the hard thermoplastic are all mixed together under high shear. The mixing temperature must be sufficient to melt the hard thermoplastic and to cause vulcanization. This takes place under high shear or "dynamic" conditions, in contrast to the static conditions of vulcanization in a typical rubber mold. When the vulcanization starts, the viscosity of the elastomer phase increases dramatically, to the point where there is a viscosity mismatch between the elastomer phase and the hard phase. This causes the elastomer phase to break up into a fine dispersion and vulcanization continues in this state. Shear must by applied until vulcanization is completed, otherwise the rubber particles can agglomerate. Small particle size is critical.

Figure 5.5 shows this effect. It is a composite stress-strain curve based on data obtained on one dynamic vulcanizate and four blends of vulcanized

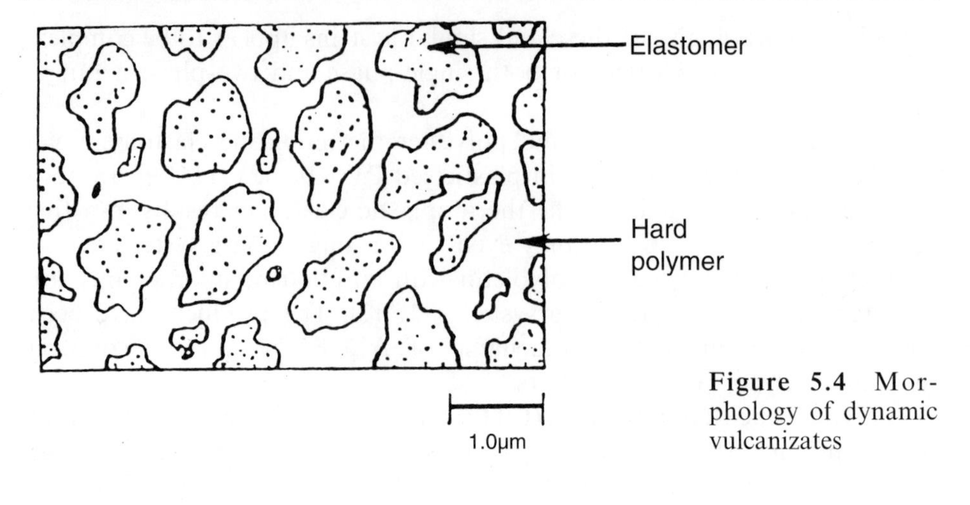

Figure 5.4 Morphology of dynamic vulcanizates

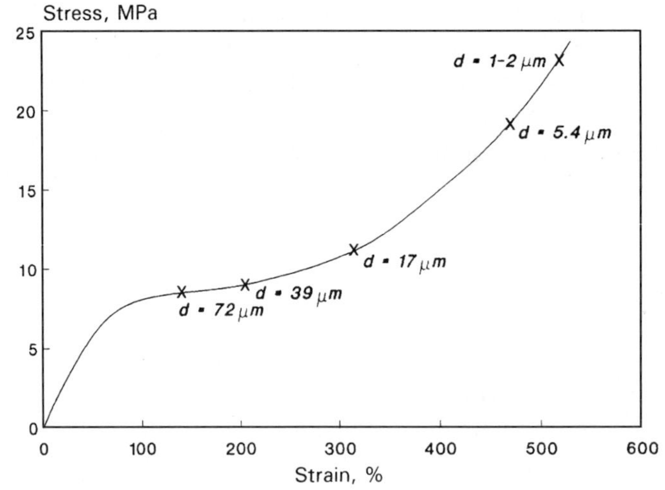

Figure 5.5 Effect of particle size on stress-strain behaviour of dynamic vulcanizates

EPDM with polypropylene. Failure points of the five compositions are marked by an X symbol. Particle sizes of the EPDM phase ranged from 72 to about 1.5 microns and are shown next to the failure points. Tensile properties improved dramatically as the particle sized decreased. For the blend with the finest particle size, both stress at break and strain at break were about three times greater than for the blend with the most coarse particle size. Thus, the energy to break (the area under the stress-strain curve) was about ten times greater for the blend with the finest particle size.

The dynamically vulcanized thermoplastic elastomers have been the subject of much research and have had many commercial applications

[1, 3, 5 to 8], probably because of the improvements these elastomers contribute to such properties as compression set and oil resistance. The authors of an authoritative survey of the subject [5] give several criteria that must be met for the end product to have satisfactory properties:

1. The surface energies of the hard phase and the elastomer phase must match.
2. The molecular weight between inter-chain entanglements in the elastomer phase must be low.
3. Crystallinity must occur in the plastic phase.
4. The elastomer phase must be able to be vulcanized at the mixing temperature.
5. Both phases must be thermally stable at the mixing temperature.

In addition to all these factors, the choice of vulcanizing agents is also important. Typical vulcanizing systems produce sulfur crosslinks. This is the process most often used to produce rubber articles from "raw" (i.e., unvulcanized) rubbers such as natural rubber, SBR, and EPDM. Similarly, sulfur-based vulcanization can be used to produced dynamic vulcanizates from polypropylene and EPDM. However, other vulcanization systems give better properties to the final product. A patent [21] describes a sulfur-free system using phenols with substituted methylol (-CH_2OH) groups. It appears that in polypropylene/EPDM dynamic vulcanizates (and probably in other dynamic vulcanizates also) the choice of the vulcanization system is a critical parameter in optimizing properties.

When compared to corresponding hard polymer/elastomer blends, dynamic vulcanizates have three significant advantages:

1. The amount of elastomer in the total blend can be increased. If a dispersion of fine particles of the vulcanized elastomer is formed during mixing, then this eliminates the possibility of a weak continuous elastomer phase. Thus, softer products with good physical properties can be obtained.
2. Because it is crosslinked, the elastomer phase cannot dissolve in oils or gasoline. This gives it improved resistance to these and similar fluids.
3. When it is stressed, an unvulcanized elastomer flows. When the stress is removed, it does not return to its original form; it develops a permanent set. In contrast, vulcanized elastomers show much less permanent set, and so dynamic vulcanizates have much better resistance to compression set.

From a commercial standpoint, the most important dynamic vulcanizates are those based on polypropylene and EPDM (EPR is not suitable because it lacks the double bonds needed for most vulcanization systems). The reasons for this are the same as the reasons for the use of similar systems to produce simple blends, that is, such factors as cost, compatibility, and service temperature range.

Although EPDM is the rubber most commonly used in polypropylene/elastomer dynamic vulcanizates, products based on natural rubber [8] and polyisobutylene [9] have also been commercialized. The first is lower in cost (although it should have reduced thermal stability), while the second has greatly reduced air permeability [5]. It should also have excellent damping and frictional properties. In both these elastomers (especially polyisobutylene), the molecular weight between chain entanglements is much higher than that of EPR see Table 3.3). According to the criteria listed above, this could give less than optimum properties to these products. However, the properties obtained in practice are apparently good enough for commercialization. One unexpected effect is that polypropylene/elastomer dynamic vulcanizates based on blends of polyisobutylene and polychloroprene have advantages over similar dynamic vulcanizates based on only one of these elastomers [22].

Almost all the above compositions are based on non-polar elastomers (polychloroprene is the exception). These are compatible with hydrocarbon oils, gasoline, and similar fluids. Thus, although the polypropylene phase in the polypropylene/elastomer dynamic vulcanizate is quite resistant to these fluids, the elastomer phase is less so. Although it is vulcanized, the elastomer phase is swollen and weakened by these fluids. This in turn swells and weakens the dynamic vulcanizate. Clearly, we would get better properties by using an oil-resistant elastomer.

A well-known one is NBR, a copolymer of acrylonitrile and butadiene. This is a polar polymer, which is why it is oil resistant, so the match between the surface energies of the hard polypropylene phase and the elastomer phase is quite poor. The solution to this problem is to use a compatibilizing agent, as shown in an idealized fashion in Fig. 5.6. The compatibitizing agent behaves similarly to a surfactant at an oil/water interface. That is, it lowers the interfacial surface tension and allows the formation of a fine dispersion of one phase in the other. The compatibilizing agent for polypropylene and NBR was prepared by grafting the two polymers together. The graft is random, that is, it is not a structure where one polymer is the backbone chain and the other is attached as side branches. Surprisingly little of this graft copolymer is needed. In a dynamic vulcanizate containing equal parts of polypropylene and NBR, clear improvements in properties were obtained when as little as 0.15% of the

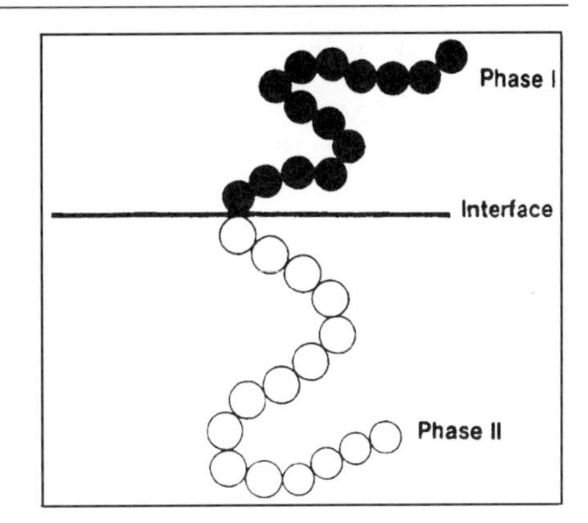

Figure 5.6 Compatilizing block copolymer molecule at an interface (idealized)

NBR was grafted onto the polypropylene. There were no significant advantages in going beyond about 2% grafting [5].

While polypropylene is an excellent choice for the hard phase, polymers with higher crystal melting points can also be used. Nylon is one such polymer and dynamic vulcanizates with NBR have been prepared and show good properties. The choice of vulcanizing system is critical. It is possible that the systems that give the best properties also graft some of the NBR onto the nylon and so form compatibilizing agents [5]. Properties of nylon/NBR dynamic vulcanizates can also be improved by adding a small amount of a functionalized polypropylene grafted with maleic anhydride. It is thought that the resulting acid groups react with the amine groups in the nylon. Similar dynamic vulcanizates based on nylon can also be made by using polyurethane rubber or poly(epichlorohydrin) rubber as the elastomer phase [5].

Another possible choice for the hard phase in dynamic vulcanizates is polybutylene terephthalate (PBT). It also has a high crystal melting point. As noted above, it has been used in both blends and dynamic vulcanizates based on EPDM as the elastomer phase [3]. Once again, a graft copolymer was used as a compatibilizing agent.

References

1 Kresge, E. N., in Chapter 5, *Thermoplastic Elastomers*, 2nd Ed., G. Holden, N.R. Legge, R.P. Quirk and H.E. Schroeder Eds. (1996), Hanser Publishers, Munich, Vienna, New York
2 Laird, J. L. (1997) *Rubber World, 217(1)* 42
3 Kresge, E. N. (1997) *Rubber World, 217(1)* 30
4 Shedd, C. D. in Chapter 3, *Handbook of Thermoplastic Elastomers*, 2nd Ed., B. M. Walker, and C. P. Rader Eds., (1998), Van Nostrand Reinhold, New York
5 Coran, A. Y. and Patel, R. P. in Chapter 7 of Reference 1

6 Rader, C. P. in Chapter 4 of Reference 4
7 Puydak, R. C., paper presented at the 2nd International Conference on Thermoplastic Elastomer Markets and Products sponsored by Schotland Business Research, Orlando, FL, March 15–17, 1989
8 Tinker, A. J., paper presented at the Symposium on Thermoplastic Elastomers sponsored by the ACS Rubber Division, Cincinnati, OH, October 18–21, 1988
9 Stockdale, M., paper presented at the Symposium on Thermoplastic Elastomers sponsored by the ACS Rubber Division, Cincinnati, OH, October 18–21, 1988
10 Tandon, P. and Stockdale, M., paper presented at the 4th International Conference on Thermoplastic Elastomer Markets and Products sponsored by Schotland Business Research, Orlando, FL, February 13–15, 1991
11 Hoffman, G. H., in Chapter 6 of Reference 1
12 Wallace, J. G. in Chapter 1 of Reference 4
13 Holden, G. in Chapter 16 of Reference 1
14 Gergen, W. P. et al. in Chapter 11 of Reference 1
15 Paul, D. R. and Barlow, J. W. (1986) *Advances in Chem. Series 211*, 3
16 Henderson, D. F. (1943) U.S. Patent 2,330,353, (to B.F. Goodrich Co.)
17a Semon, W. L. (1933) U.S. Patent 1,929,453, (to B. F. Goodrich Co.)
17b Semon, W. L. (1972) *Ann. Tech. Conf. Soc. Plas. Eng., 30* 693
18 Holden, G. in *Block and Grafts Copolymerization*, R. J. Ceresa Ed. (1973) John Wiley and Sons, New York, London
19 Adams, R. K., Hoeschele, G. K. and Witsiepe, W. K. in Chapter 8 of Reference 1
20 Goodman, I. "Polyesters", in *Kirk-Othmer Encyclopedia of Polymer Science and Engineering*, 2nd Ed, J. I. Kroschwitz, Ed. (1986), John Wiley & Sons, Inc., New York
21 Abdou-Sabet, S. and Fath, M. A. (1982) U.S. Patent 4,311,628 (to Monsanto Chemical Co.)
22 Puydak, R. C. and Hazleton, D. R. (1988) *Plast. Eng. 44,* 37

6 Graft Copolymers, Ionomers, and Core-Shell Morphologies

6.1 Importance

Previous sections dealt in some detail with the types of thermoplastic elastomers that have become commercially important. Naturally, these are the ones that attract most attention. However there are other potential routes to achieve these properties and three are described here, graft copolymers, ionomers and core-shell morphologies. They are presented in this order for two reasons. First, it appears to be the order of the degree of investigation; in other words, the most work seems to have been reported on graft copolymers and the least on thermoplastic elastomers with core-shell morphologies. Second, it also appears to be the order of the degree of atypicality of the systems. The structure and properties the graft copolymers resemble those of the styrenic block copolymers and the multi-block copolymers. On the other hand, ionomers and core-shell morphologies are very different from any of the other types of thermoplastic elastomers.

Even though these types of thermoplastic elastomers have not been commercialized, there are still valuable conclusions to be drawn from them and their properties. It is possible that this information may point the way to commercial materials from other polymers or by other methods. Sometimes the knowledge of what does not work, and why, can give us insight into products that do work.

6.2 Graft Copolymers

6.2.1 Structure

The composition of a number of graft copolymers that are thermoplastic elastomers is shown in Table 6.1. The generalized structure is represented as:

Table 6.1 Thermoplastic Elastomers Based on Graft Copolymers[a]

Hard Pendant Segment, *H*	Soft or Elastomeric Backbone Segment, *E*	Refs
Polystyrene & Poly(α-methyl styrene)	Polybutadiene	[1, 2]
Polyindene	Polybutadiene	[3]
Polystyrene & Poly(α-methyl styrene)	Poly(ethylene-co-propylene)	[4–6]
Polyindene	Poly(ethylene-co-propylene)	[3]
Polyindene	Polyisobutylene	[3, 7]
Polystyrene & Poly(α-methyl styrene)	Polyisobutylene	[7–9]
Poly(phenylene ether)	Polyisobutylene	[8]
Polyacenapthylene	Polyisobutylene	[7]
Poly (para chlorostyrene)	Polyisobutylene	[10]
Polystyrene	Chlorosulfonated polyethylene	[11]
Poly(α-methyl styrene)	Polychloroprene	[9]
Polystyrene	Poly(butyl or ethyl-co-butyl) acrylates	[12]
Poly(methylmethacrylate)	Polybutylacrylate	[13]
Polypivalolactone	Poly(ethylene-co-propylene)	[14, 15]
Polypivalolactone	Polyisobutylene	[16]

[a]For more detailed information, see References [12 and 18].

$$E \ldots E\text{-}\begin{pmatrix}\text{-}E\text{-} \\ | \\ H\end{pmatrix}_n\text{-}E\text{-}E\text{-}E \ldots E$$

This represents a polymer where each elastomeric poly*E* backbone chain has (on average) n random grafts of pendant hard *H* segments. This is sometimes called a "comb" structure. If the molecular weight of the *H* segments is sufficiently high, they phase separate and form an interconnected network similar to that formed by the physical crosslinking of linear styrenic triblock copolymers (Fig. 3.1).

6.2.2 Chain Statistics

There are several critical differences between the structure of linear styrenic triblock copolymers and that of graft copolymers. First, triblock copolymers have exactly two hard segments per molecule, one at each end of the elastomer chain. In contrast, the number of hard segments per molecule in a graft copolymer is under no such restriction. In principle, any number of hard segments can be positioned randomly along each elastomer chain. In practice, the number can be varied over a very wide range by changing the conditions of manufacture. The average number of hard segments per elastomer chain is denoted as n. Note the use of the word "average." Even if the molecular weights of all the elastomer chains are similar (which in general, they are not) the number of hard *H* segments per elastomer chains is a statistical distribution. Thus, some elastomer chains have no *H* segments, some have one, some have two, etc. Elastomer chains that have fewer than two *H* segments are not elastically effective (see Section 3.1). The statistical treatment of this is quite complex [8, 17]. Even this is not the whole story.

Consider a graft copolymer with the structure represented below:

```
E-E-E-E-E-E-E-E-E-E-E-E-E-E-E-E-E-E-E-E-E
         |                   |
         H                   H
```

This represents a graft copolymer with two *H* segments. Such a graft copolymer can take part in a physically crosslinked structure such as shown in Fig. 3.1 for linear triblock copolymers. But there is an important difference. The graft copolymer has some fraction of its elastomer chains in "dangling ends." These are the part of the elastomer chain that is not between two *H* segments . This part of the elastomer chain is not elastically effective. In the above example, about 2/3 of the elastomer chain is ineffective. This is the statistical expectation for a long polymer chain with two grafts. Although the elastically effective fraction increases as n is increased, there is always some elastically ineffective material present. In contrast, in a linear triblock copolymer, the entire elastomer chain is elastically effective.

The two effects (the fraction of elastomer chains with less than two *H* segments and the fraction of elastomer chains in "dangling ends") can be combined (Fig. 6.1). The results suggest that graft copolymers of this type have reasonable properties only if n is significantly greater than 2. Values from 4 [8] to as high as 10 [18] have been suggested.

6.2.3 Molecular Parameters

It is also necessary that the molecular weight of the *H* segment and the incompatibility between the backbone *E* chain and the pendant *H* segments

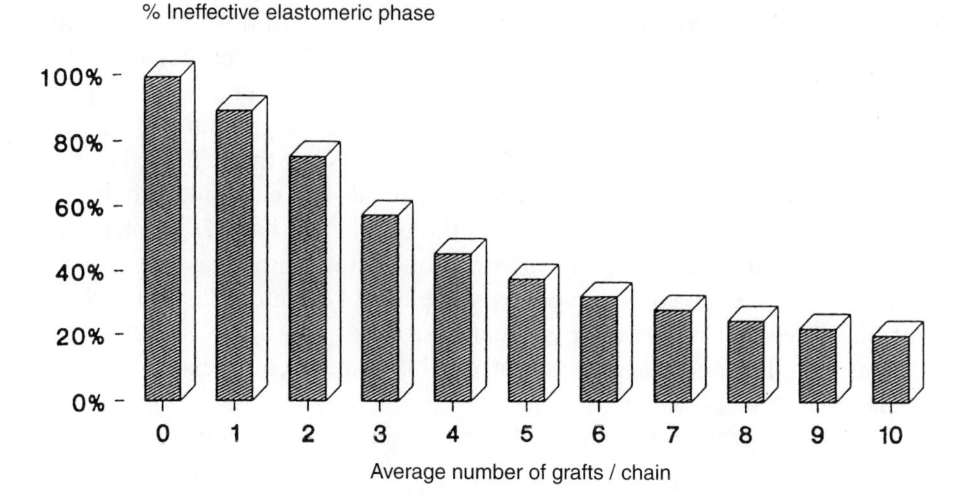

Figure 6.1 Fraction of Ineffective Elastomer Phase in Graft Copolymers

should be high enough to give good phase separation. The required molecular weight depends on the structural differences between the polymer forming the *E* backbone chain and that forming the pendant *H* segments. For S-B-S polymers, this critical molecular weight is about 7,000 (see Section 3.1.5). In practice, the minimum value required to give acceptable properties is somewhat higher, approaching 10,000. This higher value tends to favor a more complete phase separation and reduces phase mixing at the interface. A study of polyisobutylene/polystyrene graft copolymers concluded that the minimum molecular weight for phase separation was in the range of 3,000 to 4,000, and the minimum molecular weight for effective domain formation was about 6,000 [8]. These lower values than those found for the S-B-S block copolymers are probably the result of the greater degree of incompatibility between polyisobutylene and polystyrene as compared to that between polybutadiene and polystyrene.

As would be expected, the properties of these graft copolymers depend on the ratio of the soft phase to the elastomer phase. Figure 6.2 shows some stress-strain curves for a series of graft copolymers with polystyrene segments grafted onto ethylene-propylene rubber backbones [18]. As the polystyrene content of the graft copolymer increases, it changes from a soft to a hard elastomeric material and then, at about 56% polystyrene content, to a hard, leathery material with a yield point. This behavior is obviously

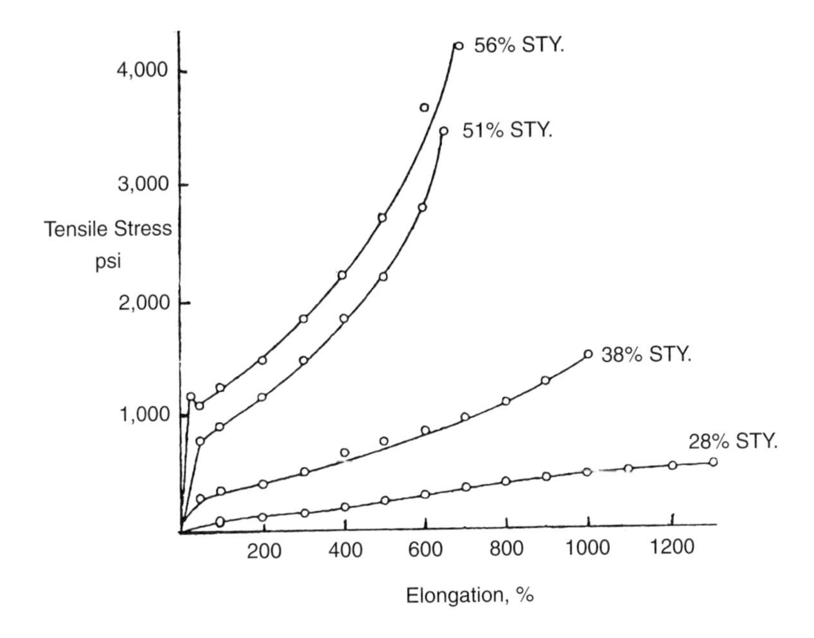

Figure 6.2 Stress-Strain Properties of EPR-Polystyrene Graft Copolymers at Varying Polystyrene Contents

similar to that of S-B-S block copolymers with varying polystyrene contents (Fig. 3.6). Similar behavior is described in recent articles on graft copolymers and their properties [12, 18, 19]. However, none of the graft copolymers materials with about 30% hard phase content display the combination of high tensile strengths (>25 MPa) and high elongations (>500%) typically observed in analogous styrenic block copolymers (see Fig. 3.9). These decreases in properties found in the graft copolymers are probably the result of elastically ineffective material (i.e., "dangling ends," ungrafted and monografted elastomer), as described earlier.

6.2.4 Production

Almost all these graft copolymers are produced by basically the same technique. The process starts with an elastomeric polymer that has active sites along its chain. There are several possibilities:

Elastomer	**Active Site**
Polybutadiene	In-chain double bonds
Poly(ethylene-co-propylene) EPDM	Pendant double bonds
Poly(ethylene-co-propylene)	Pendant halogen atoms
Polyisobutylene	Pendant halogen atoms
Chlorosulfonated polyethylene	Pendant chlorine atom
Polychloroprene	Pendant chlorine atoms

There are then two possibilities. In one, the active sites initiate polymerization of a monomer that becomes the hard phase. An alternative scheme uses low molecular weight polymers (e.g., 10,000 molecular weight polystyrene) that have reactive end groups. These attach to the active sites on the elastomeric backbone to form the hard phase. In both cases, the hard polymer groups arc pendant to the elastomer chain.

Another technique uses "macromonomers" [12]. In this case, polymer moities (say, polyA) with a polymerizable group are copolymerized with a monomer (say, B). The result is essentially a random copolymer of polyA and B. It differs from a conventional random copolymer of A and B in that instead of A being a monomer unit (e.g., styrene), it is a polymer (e.g., polystyrene). If B is a monomer that polymerizes to form an elastomer (e.g., butadiene), the result is a graft copolymer with a hard and soft phase. The hard phase consists of pendant groups of polyA. These are attached to an elastomeric backbone chain of polyB.

6.2.5 Properties

In all these cases, the number of these pendant *H* groups that attach to each elastomer molecule is a statistical distribution; in other words, we can only speak of the average number (n) of pendant groups per elastomer molecule. For satisfactory properties, n must be at least 4. Thus, if the hard segment molecular weights and proportions are the same, a graft polymer with $n = 4$ has twice the molecular weight of the corresponding H-E-H triblock copolymer. If $n > 4$, the disparity is still greater. For this reason, at equal hard segment molecular weights, graft copolymers have higher viscosities than equivalent triblock copolymers, which should make them more difficult to process. Additionally, they contain many dangling ends (see above).

However, the potential to use pre-polymerized elastomer chains is an advantage. These elastomer chains can be resistant to degradation (e.g., poly(ethylene-co-propylene) [4 to 6] or polyisobutylene [3, 7 to 10, 16], and they can combine this stability with improved oil resistance (e.g., chlorosulfonated polyethylene) [11]. The glass transition temperature of the hard segments can be very high (e.g., PPE, polyphenylene ether [8], for which $T_g = 205\,^{\circ}C$). Alternatively, crystalline polymers can be used. Grafts of polypivalolactone (-[-CH_2-$C(CH_3)_2$-COO-]$_n$-) on poly(ethylene-co-propylene) or polyisobutylene had good mechanical properties with crystal melting temperatures of up to 200 °C [14 to 16]. These polymers are noteworthy because the difference in solubility parameter between the hard and soft phases is extremely high, leading to a high degree of phase separation in the melt.

Although graft copolymers may usually have somewhat lower strength properties than block copolymers, this is not necessarily a drawback. At first glance, a polymer that is strong appears to have an inherent advantage. As discussed in more detail in the next chapter, this strength allows such a polymer to be compounded with lower cost ingredients. But sometimes (specifically in sealants and adhesives), high strength can be a disadvantage. Because of this, styrenic block copolymers with up to 80% diblock content are commercially produced and are blended with the corresponding triblock copolymers to reduce the strength of the final product (see Section 7.12).

6.3 Ionomers

The structure of ionomers that are thermoplastic elastomers may be written as:

$$\begin{array}{c} E \ldots E\text{-}E\text{-}E^{-}\text{-}E\text{-}E \ldots E \\ (M^{+})_n \end{array}$$

This represents an elastomer molecule that has acidic groups (E^-) polymerized into the backbone chain. These are neutralized by metallic counterions (M^+). These counterions associate together to form ionic clusters that can behave like the hard polymer domains in block and graft copolymers; in other words, they tie the elastomer chains together in a physically vulcanized network. The concept is based on the conventional ionomer system that has been in commercial use for many years [20]. This system gives flexible rather than elastomeric products because the backbone chain is polyethylene. Replacement of the polyethylene by an elastomeric polymer is an obvious extension. Many different acidic groups have been investigated, including carboxylates and sulfonates [21].

There are many elastomers that could be used in such a system. Some examples are polyisobutylene, EPDM and polypentenamer

(-[-$(CH_2)_3$-CH=CH-]$_n$-). The system that appears to have most potential is sulfonated EPDM with zinc counterions. This is blended with zinc stearate, which plasticizes the product. The elastomer is then further compounded with hydrocarbon oils, fillers, and polyolefins such as polyethylene or polypropylene. Note the similarity to the compounding of styrenic block copolymers, particularly S-EB-S. Thermoplastic elastomers with hardness as low as 70 Shore A can be produced [21], but have not been commercialized.

6.4 Core-Shell Morphologies

Core-shell mophologies are multi-component polymers produced as very small, spherical particles. The polymer composition at the center of the sphere (the core) differs from that at the surface (the shell). Latex polymerization is the obvious way to produce such a material. One example [22] has a rigid core (a styrene-acrylonitrile copolymer) and an elastomeric shell (polybutylacrylate). A transition zone separates the core from the shell and is an essential component of the system.

Another possibility is the opposite morphology, a rigid shell and an elastomeric core. If the polymer used as the shell is compatible (in the sense of forming a good mechanical bond) with a hard thermoplastic of some type, it allows the elastomer phase to act as an impact modifier for this hard thermoplastic. The process could be carried further by cross-linking the elastomer phase during polymerization. If the particle size of the core-shell material is small enough and if sufficient of it is added to the hard thermoplastic, the product should be similar to the thermoplastic elastomers produced by dynamic vulcanization (see Section 5.2).

References

1 Kennedy, J. P. and Delvaux, J. M. (1981) *Adv. Polym. Sci. 38*, 141
2 Ambrose, R. and Newell, J. J. (1979) *J. Polym. Sci. Polym. Chem. Ed. 17*, 2129
3 Sigwalt, P., Polton, A. and Miskovic, M. (1976) *J. Polym. Sci. Symp. No. 56*, 13
4 Kennedy, J. P. and Smith, R. R. In *Recent Advances in Polymer Blends, Grafts and Blocks* L. H. Sperling Ed. (1974) Plenum Press, New York and London
5 Smith, R. R. (1984) Ph. D. Thesis, The University of Akron
6 Gadkari, A. and Farona, M. (1987) *Polym. Bull 17*, 229
7a Kennedy, J. P. and Charles, J. J. *J. Appl. Polym. Sci. Appl. Polym. Symp. 30*, 119
7b Charles, J. J. (1983) Ph. D. Thesis, The University of Akron
8 Tse, M. F., Dias, A. J. and Wang, H.-C. (1999) *Rubber Chem. Technol. 71*, 803
9 Kennedy, J. P. and Guhaniyogi, S. C. (1982) *J. Macromol. Sci. Chem. A18*, 103

10 Kennedy, J. P. and Baldwin, F. B. (1968) Belgian Patent 701,850
11 Kennedy, J. P. and Metzler, D. M. *J. Appl. Polym. Sci. Appl. Polym. Symp. 30*, 105
12a Quirk, R. P., Brittain, W. J. and Schultz, G. O. in Chapter 14. *Thermoplastic Elastomers*, 2nd Ed., G. Holden, N. R. Legg, R. P. Quirk and H. E. Schroeder Eds., (1996), Hanser Publishers, Munich, Vienna, New York
12b Schultz, G. O. and Milkovich, R. (1982) *J. Appl. Polym. Sci. 27*, 4473
13 Xie, H. and Zhoui, S. (1990) *J. Macromol. Sci. Chem. A27,* 491
14 Sundet, S. A. et al. (1976) *Macromolecules 9*, 371
15 Thamm, R. C. and Buck, W. H. (1978) *J. Polymer Sci. Polym. Chem. Ed. 16*, 539
16 Harria, J. F. Jr., and Sharkey, W. H. (1986) *Macromolecules 19(12)*, 2903
17 Tung, L. H. and Wiley, R. M. (1973) *J. Polym. Sci., Polym. Phys. Ed. 11*, 1413
18 Kennedy, J. P. in Chapter 13 of Reference 12a
19 Kresge, E. N. in Chapter 5 of Reference 12a
20 Rees, R. W. in Chapter. 10A of Reference 12a
21 McKnight, W. J. and Lundberg, R. D. in Chapter 10B of Reference 12a
22 Falk, J. C. and Van Beck, D. A. (1984) U. S. Patent 4,473,679, (to Borg-Warner Chemicals, Inc.)

7 Commercial Applications of Thermoplastic Elastomers

Since their commercial introduction, many types (and grades within types) of thermoplastic elastomers have been produced to meet specific end use requirements. About 50 manufacturers have entered the field.

7.1 Styrenic Block Copolymers

Like most conventional vulcanized rubbers – and unlike most thermoplastics – styrenic block copolymers are never used commercially as pure materials. To achieve the particular requirements for each end use, they are compounded with other polymers, oils, resins, fillers, etc. Usually, the final products contain less than 50% of the block copolymer. Thus, a study of their end uses is in effect a study of how they are blended to achieve the properties needed for the particular application.

Before discussing the end uses in detail, it is important to consider how the added materials are distributed with respect to the two phases in the block copolymer. For any additive there are four possibilities:

1. It can go into the polystyrene phase. In this case, the additive increases the relative volume of the polystyrene phase and so makes the product harder. The glass transition temperature of the additive should be similar to or greater than that of polystyrene (100 °C). If not, its addition reduces the high temperature performance of the final product.
2. It can go into the elastomer phase. Conversely, in this case, the additive decreases the relative volume of the polystyrene phase and so makes the product softer. The addition also changes the glass transition temperature of the elastomer phase. This, in turn, affects such end use properties as tack and low temprature flexibility.
3. It can form a separate phase. Unless the molecular weight of the additive is substantially less than that of either segment in the block

copolymer, this is the most likely outcome (see Section 3.1.6). Thus, only low molecular weight resins and oils are compatible with either of the existing two phases, and polymeric materials usually form a separate third phase. This polymeric third phase is usually co-continuous with the block copolymer and so confers some of its own characteristic properties on those of the final blend.

4. It can go into both phases. This is usually avoided because it reduces the degree of separation of the two phases and so, weakens the product.

This ability to be blended with so many different materials gives styrenic block copolymers an exceptionally wide range of end uses. Some examples are reviewed in the following sections.

7.1.1 Styrenic Block Copolymers as Replacements for Vulcanized Rubber

In this end use, the products are usually manufactured by machinery originally developed to process conventional thermoplastics. Examples are injection molding, blow molding, blown film and profile extrusions. S-I-S block copolymers have not been used very much as replacements for vulconized rubber, but there are many applications for compounded products based on S-B-S and S-EB-S block copolymers. A list of some compounding ingredients and their effects on the properties of the compounds is given in Table 7.1. Products can be produced that are as soft as 5 Shore A or as hard as 55 Shore D. Large amounts of these compounding ingredients can be added. Often the final products contain only about 25% of the S-B-S or S-EB-S block copolymers. From an

Table 7.1 Compounding Styrenic Block Copolymers

Component	Effect on....				
	Hardness	Processability	Oil/Solvent Resistance	Cost	Other
Oils	Decreases	Improves	–	Decreases	–
Polystyrene	Increases	Improves	–	Decreases	–
Polypropylene	Increases	Improves, especially with S-EB-S	Improves	Decreases	Improves high temp. resistance
Filler	Slight increase	Slight improvement	–	Decreases	Increases density

economic point of view, this is most important. For example, it enables compounds based on S-EB-S block copolymers to compete with those based on polypropylene/EPDM or EPR blends, although the S-EB-S alone is several times more expensive than the polypropylene, the EPDM, or the EPR.

Polystyrene is often used as a compounding ingredient for S-B-S block copolymers. It acts as a processing aid and makes the products stiffer. Mineral oils are also processing aids, but make the products softer. Naphthenic oils are preferred. Oils with high aromatic contents should be avoided, since they plasticize the polystyrene domains. Inert fillers such as whiting, talc, and clays can also be used in these compounds. They have only a relatively small effect on physical properties, but reduce cost. Frequently, up to 200 parts of filler per 100 parts of the block copolymer are used in compounds intended for footwear applications. Reinforcing fillers such as carbon black are not required. In fact, large quantities of such fillers make the final product stiff and difficult to process.

S-EB-S block copolymers can be compounded in a similar manner to their S-B-S analogs. One important difference is that for S-EB-S block copolymers, polypropylene is a preferred additive, acting in two different ways to improve the properties of the compounds. First, it gives the compounds better processability. Second, when the compounds are processed under high shear and then quickly cooled, (e.g., in injection molding or extrusion), the polypropylene and the S-EB-S/oil mixture form two co-continuous phases. Polypropylene is insoluble at service temperatures and has a relatively high crystal melting point of about 165 °C. Thus, the continuous polypropylene phase significantly improves both the solvent resistance and upper service temperature of these compounds.

Another advantage of S-EB-S polymers is that because of their lower midsegment solubility parameter, they are very compatible with mineral oils. Large amounts of these oils can be added without bleedout, allowing very soft compounds to be produced. Paraffinic oils are preferred because they are more compatible with the EB center segments than the naphthenic oils used with similar S-B-S based compounds. Again, oils with high aromatic contents should be avoided.

Blends of S-EB-S block copolymers with mineral oils and polypropylene are transparent because the refractive index of an S-EB-S / oil mixture almost exactly matches that of crystalline polypropylene. Surprisingly, even oils that are stable to UV radiation reduce the stability of the blends, but these effects can be minimized by using UV stabilizers and absorptive or reflective pigments. Blends with silicone oils are used in some medical applications [1]. The same inert fillers used in the S-B-S based compounds can also be used with the S-EB-S analogs. In addition, barium or strontium sulfates can be used to produce very high density compounds used for sound deadening applications. In another development, fire retardants can be added to produce compounds that meet the requirements of many current regulations.

Compounds based on both S-B-S and S-EB-S thermoplastic elastomers require some protection against oxidative degradation and in some cases, against sunlight also, depending on their end use. Hindered phenols are effective antioxidants and are often used in combination with thiodipropionate synergists. Benzotriazoles are effective UV stabilizers and are often used in combination with hindered amines. If transparency is not essential in the end product, titanium dioxide or carbon black pigments provide very effective protection against sunlight.

Compounding techniques are relatively simple and standard. There is one important generalization: the processing equipment should be heated to temperatures at least 20 °C above the upper service temperature of the block copolymer (see Table 2.1) or the melting point of the polymeric additive, whichever is the greater. The use of cold mills, etc., can result in polymer breakdown, which is not only unnecessary, but also detrimental to the properties of the final product.

Unfilled or lightly filled compounds can be made on a single screw extruder fitted with a mixing screw. The length/diameter ratio should be at least 24:1. If large amounts of fillers or fire retardants are to be added, these can be dispersed on either a twin screw extruder or a closed intensive mixer that discharges into an extruder. After extrusion, the compound is fed to a pelletizer, either a strand cutting or an underwater face cutting system. If a strand cutting system is used, it is important to remember that rubbery compounds must be cut rather than shattered. Thus, the blades of the cutter should be sharp and the clearance between the fixed and the rotating blades should be minimized. With this type of pelletizer, the strands must be thoroughly cooled before they enter the cutter. A chilled water bath can be used to increase production rates.

Trial batches or small scale production runs can be made using a batch type, closed intensive mixer. Mixing times of about five minutes are usually adequate. After this mixing, the hot product is passed to a heated two-roll mill. When the product has banded, it is cut off, allowed to cool, and finally granulated. In this case also, the granulator blades must be sharp and clearances minimized.

S-B-S and S-EB-S block copolymers with high molecular weight, polystyrene segments and/or high styrene contents are very difficult to manufacture and process as pure materials. Oiled versions with oil contents from about 25 to 50% are commercially available. These process much more easily and of course, more oil can be added during mixing.

Many end users prefer to buy pre-compounded products and numerous grades have been developed for various specialized end uses. These include milk tubing, shoe soles, sound deadening materials, wire insulation, and flexible automotive parts. After priming, the parts can be coated with flexible paints. If a compound contains a particular homopolymer (e.g., polystyrene or polypropylene), it adheres well when insert-molded or extruded against the same pure homopolymer or against other compounds containing it. This allows the production of parts with rigid frameworks supporting soft, flexible outer surfaces [2].

The physical properties of compounded thermoplastic elastomers based on both S-B-S and S-EB-S block copolymers are sensitive both to processing conditions and to processing equipment. Thus, it is most important to make test samples under conditions and on equipment similar to those that will be used in production. Misleading results will be obtained if, for example, prototype parts or test pieces are compression molded when the actual products will be made by extrusion or injection molding.

Conditions for processing these compounds by injection molding, blow molding, extrusion, etc. have been discussed in some detail in recent literature [3]. As a generalization, compounds based on S-B-S block copolymers are processed under conditions suitable for polystyrene while those based on S-EB-S block copolymers are processed under conditions suitable for polypropylene. These compounds especially the softer versions, usually have relatively high surface friction. Ejection of the molded parts can be difficult, especially with softer products. Use of a release agent or a suitable coating on the mold makes for easier ejection. If possible, the mold should be designed so that ejection can be air assisted. Tapering the sides of the mold is also helpful, as is the use of stripper rings. The use of small diameter ejection pins should be avoided because they often deform the molded part rather than eject it.

Ground scrap from molding is reusable and is usually blended with virgin product. Grinding is relatively easy if the conditions required for successful pelletization are met, i.e., if the grinder blades are kept sharp and the clearances minimized.

Various compounds have been developed to allow the production of blown and extruded (slot cast) film, including heat shrinkable films [4]. These are based on both S-B-S and S-EB-S block copolymers and can be very soft and flexible. They also exhibit low hysteresis and low tensile set. A significant advantage is that they can be used in contact with skin and in contact with certain foods

One unusual application is the use of solutions of S-EB-S/oil blends to replace natural rubber latex in the manufacture of articles such as surgeon's gloves [5]. These blends have two advantages. First, they are more resistant to attack by oxygen or ozone. Second, natural rubber latex contains proteins that produce dangerous allergic reactions in some patients. The saturated elastomer phase and extreme purity of the S-EB-S avoids both these problems.

7.1.2 Styrenic Block Copolymers as Adhesives, Sealants, and Coatings

These are very important applications for styrenic block copolymers and probably the fastest growing. Again, the products are always compounded and the subject has been extensively covered [6, 7]. As already mentioned, the effects of the various compounding ingredients depend on the region of the phase structure with which they associate. Because there are three different elastomers used in these block copolymers, each has particular resins and/or oils with which it is most compatible. Table 7.2 gives details of the various resins and oils suitable for use with each of these elastomers and also with the polystyrene phase. Ingredients that go into both phases are usually avoided, because they make the phases more compatible with each other, and thus, make the product weaker. Polymers that go into neither phase (e.g., polypropylene) are used in some hot melt applications where

Table 7.2 Resins Etc. Used to Formulate Adhesives, Sealants, etc. from Styrenic Block Copolymers

Resin Type	Segment Compatibility*
Polymerized C5 Resins (Synthetic Polyterpenes)	I
Hydrogenated Rosin Esters	B
Saturated Hydrocarbon Resins	EB
Naphthenic Oils	I, B
Paraffinic Oils	EB
Low Molecular Weight Polybutenes	EB
Aromatic Resins	S

* I - Compatible with polyisoprene segments
B - Compatible with polybutadiene segments
EB - Compatible with poly(ethylene-butylene) segments
S - Compatible with polystyrene segments

they stiffen the products and increase the upper service temperature. Fillers can also be added to reduce cost.

The products can be applied either as solutions or as hot melts. Because the molecular weights of these block copolymers are relatively low (typically less that 150,000), the solutions can be made at high solids content. Some details of the behavior in various solvents are given in the literature [8]. Hot melt application is often preferred because it avoids many problems associated with solvents, such as flammability, toxicity, and air pollution. In this case, the molten resins and/or oils can be regarded as replacing the solvents as well as modifying the properties of the end product. Application rates of hot melt products are usually faster than those of solvent-based analogues because the time required for a product to cool is much less than the time required for a solvent to evaporate. Some details about the various adhesive applications follows:

7.1.2.1 Pressure Sensitive Adhesives

This is probably the largest single end use for styrenic block copolymers. The products are usually applied at hot melts, although solvent application is possible and takes advantage of the low solution viscosity of these polymers. The end uses include various kinds of tapes and labels, as well as adhesive fasteners such as diaper tabs. The visco-elastic mechanism by which the resins and the thermoplastic elastomer combine to give tacky products has been described in the literature [9 to 11].

According to this theory, the resins have two functions. First, they mix with the elastomer phase in the thermoplastic elastomer and so, soften the product. This softening allows the adhesive to conform to the substrate. This is considered as the "bonding" stage of adhesion and is relatively slow. Next is the removal of the adhesive from the substrate, the "disbonding" stage of adhesion, which occurs more quickly. Here, the function of the resins is to adjust the glass transition temperature of the elastomer phase so that the adhesive stiffens up during this fast stage and resists removal from the substrate.

Because a soft product is necessary to form the adhesive bond, softer block copolymers are preferred as the materials from which to formulate pressure sensitive adhesives. These soft block copolymers usually have low polystyrene contents and a relatively high content of diblock (i.e., S-I, S-B or S-EB). This diblock content is non-load-bearing. It, and the resins in the elastomer phase, weaken the adhesive. However, this weakening can be tolerated so long as it is not carried to the point where it causes cohesive failure of the adhesive during service.

Products have recently been developed that can be crosslinked by radiation after their application to the tape [12]. This process gives the adhesive improved solvent resistance, which is important for applications such as masking tapes.

7.1.2.2 Assembly Adhesives

S-B-S and S-EB-S block copolymers are often used in this application and again, hot melt application is more usual than application from solution. Tack is not important (it may even be undesirable) and so, harder products are satisfactory. These adhesives are usually formulated to contain two types of resins, those that are compatible with the polystyrene phase and other resins (and possibly oils) that are compatible with the elastomer phase. The relative proportions of the two types of resin determines the hardness of the adhesive, while the total amount of resin added determines the viscosity of the final product.

7.1.2.3 Sealants

This application mostly uses S-EB-S block copolymers. Both hot melt and solvent-based applications are important. Hot melt sealants are applied in a variety of manufacturing processes, frequently robotically. They can also be processed as foamed products. Often they are used as formed-in-place gaskets. In contrast, the solvent-based products are limited to the building industry, where they are applied on site, both during the initial construction and during subsequent maintenance and repair.

Diblock polymers are often part of both hot melt and solvent-based sealants, because they reduce the viscosities of the hot melt products or allow the solvent-based products to be formulated at higher solids content. The diblock polymers also reduce the strength of the sealant to the point where it fails cohesively during peel, which is a requirement in many sealant specifications. Again, both types of resins are normally used and oils are frequently a part of the composition. Unless clarity is a requirement, large amounts of fillers, such as calcium carbonates, can also be added. Suggested starting formulations for both hot melt and solvent based sealants [13] have been published.

7.1.2.4 Coatings

Chemical milling of metals is an important application for coatings based on styrenic block copolymers. In this process, a protective film is first applied to the whole surface of the metal sheet and then selectively peeled away from areas from which metal is to be removed. The sheet is then immersed in an etchant bath, which dissolves away the unprotected areas. The two most common metals processed in this way are aluminum and titanium. Aluminum is etched under alkaline conditions and can be protected by coatings based on S-B-S block copolymers. Titanium, however, is etched by strongly oxidizing acids and so coatings based on S-EB-S block copolymers must be used. Both types of protective coatings are probably formulated with the usual resins, fillers, etc. but details of the compositions have not been published.

7.1.2.5 Oil Gels

As noted previously, styrenic block copolymers are very compatible with mineral oils. Blends with as little as 5% of an S-EB-S block copolymer (the remainder being 90% mineral oil and 5% wax) have been described for use as cable filling compounds [14, 15]. These fill the voids in "bundled" telephone cables and prevent water seepage. Another potential application is toys, hand exercising grips, etc [16].

7.1.3 Styrenic Block Copolymers in Blends with Thermoplastics or Other Polymeric Materials

Styrenic block copolymers are technologically compatible with a surprisingly wide range of other polymeric materials, i.e., the blends show improved properties when compared with the original polymers. Impact strength usually is the most obvious improvement, but others include tear strength, stress crack resistance, low temperature flexibility, and elongation. Thermoplastic and thermoset polymers can be modified in this way, as can asphalts and waxes.

7.1.3.1 Thermoplastics

Thermoplastic elastomers have several advantages in this application. The other elastomers that can be blended with thermoplastics to improve their impact resistance (e.g., SBR, EPDM and EPR) can normally be used only in the unvulcanized state; vulcanized products usually cannot be dispersed. An exception is the process of vulcanization during dispersion (see Section 5.2). Because these unvulcanized elastomers are soft and weak, they reduce the strength of the blends and so, only limited amounts can be added. However, thermoplastic elastomers are much stronger, even though they are unvulcanized, and so, there is less limitation on the amounts that can be added. Blending is usually carried out in the processing equipment (injection molders, extruders, etc.), especially if a thermoplastic elastomer with low viscosity is used. The thermoplastic elastomers form a separate phase and so do not change the T_g or T_m of the thermoplastic into which they are blended. Thus, these blends maintain the upper service temperature of the original thermoplastic.

Three large volume thermoplastics, polystyrene, polypropylene, and polyethylene (both high and low density), can be modified using styrenic block copolymers [17, 18]. Because of their lower price, S-B-S block copolymers are often used with these thermoplastics. In polystyrene, there are two important applications: upgrading high impact polystyrene to a super high impact product and restoring the impact resistance which is lost when flame retardants are mixed into high impact polystyrene. Polypropylene has very poor impact resistance at low temperatures. This also can be improved by adding styrenic block copolymers.

The impact improvement of any polymer is usually accompanied by a loss in clarity because the added polymer forms a separate phase with a different refractive index. However, blends of S-EB-S with polypropylene are about as transparent as the pure polypropylene, probably because of a match in refractive indices. Linear low density polyethylene (LLDPE)/S-EB-S mixtures can be used instead of the pure S-EB-S and blends of these two polymers with polypropylene also retain the clarity of the pure polypropylene [19]. Blends with polyethylene are mostly used to make blown film, where they have improved impact resistance and tear strength, especially in the seal area. Styrenic block copolymers are also blended with engineering thermoplastics such as poly(phenylene oxide) [20] and polycarbonate [21]. S-EB-S block copolymers are often used in these blends because of their better stability. In more polar thermoplastics such as polyamides and polyesters, maleated S-EB-S polymers are recommended as impact modifiers [18].

A slightly different application is the use of styrenic block copolymers to make useful blends from otherwise incompatible thermoplastics. Polystyrene, for example, is completely incompatible with polyethylene or polypropylene, and blends of this type form a two-phase system with virtually no adhesion between the phases. Thus, when articles made from them are stressed, cracks easily develop along the phase boundaries and the products fail at low elongations. Addition of a low molecular weight S-B-S or S-EB-S block copolymer changes this behavior and converts the blends to more ductile materials [18]. Similar results were reported on blends of polyphenylene oxide) with polypropylene compatibilized with S-EB-S and S-EP, and on polyamides compatibilized with polyolefins using functionalized S-EB-S copolymers. In another example, a higher molecular weight S-EB-S was used in a blend with a polycarbonate and a polypropylene. Compared with the pure polycarbonate, the blend had almost the same upper service temperature combined with improved solvent resistance, lower cost, and lower density [22].

7.1.3.2 Thermosets

Sheet molding compounds (SMC) are thermoset compositions containing unsaturated polyesters, styrene monomer, chopped fiberglass, and fillers. They are cured to give rigid parts that are often used in automobile exteriors. Special types of styrenic block copolymers have been developed as modifiers for these compositions and give the final products improved surface appearance and better impact resistance [23, 24].

In an entirely different application, thermoplastic elastomers are blended with silicone rubbers containing either vinyl or silicon hydride functional groups. The silicone rubbers containing the vinyl groups are pelletized separately from those containing the silicon hydride groups. When melted and mixed together in the processing equipment, the two groups react under the influence of a platinum catalyst to form an interpenetrating network of the silicone rubber and the thermoplastic elastomer. The products are useful in medical applications. This end use was originally developed using polyurethane/elastomer block copolymers. It has now been extended to other thermoplastic elastomers including S-EB-S and polyester elastomer block copolymers [25].

7.1.3.3 Asphalt Blends

The block copolymer content of these blends is usually less than 20%; even as little as 3% can significantly change the properties of asphalts. The styrenic block copolymers make the blends more flexible, especially at low temperatures, and increase their softening point. They decrease the penetration and reduce the tendency to flow at high service temperatures, such as those encountered in roofing and paving applications. They also increase the stiffness, tensile strength, ductility, and elastic recovery of the final products. Melt viscosities remain relatively low and so the blends are still easy to apply. As the polymer concentration is increased to about 5%, an interconnected polymer network is formed. At this point the nature of the blend changes from an asphalt modified by a polymer to a polymer extended with an asphalt. It is important to choose the correct grade of asphalt; those with low asphaltenes content and/or high aromaticity in the maltene fraction usually give the best results [26]. Applications include road surface dressings such as chip seals, which are applied to hold the aggregate in place when a road is resurfaced; slurry seals; asphalt concrete; which is a mixture of asphalt and aggregate used in road surfaces; road crack sealants; roofing; and other waterproofing applications [27 to 29]. Their lower cost makes S-B-S block copolymers the first choice for this application, but in roofing and paving applications, the S-EB-S block copolymers are also used because of their better UV, oxidative, and thermal stability.

7.1.3.4 Wax Blends

S-I-S and S-B-S block copolymers can be used in blends with waxes. Because of their limited compatibility, they usually require the addition of resins. S-EB-S block copolymers are more compatible. The products are used as flexible coatings for paper products and can be applied by curtain coaters [30].

7.2 Multi-Block Copolymers

The structure of these block copolymers is an alternating series of polymeric segments. One type of segment forms a hard phase (usually crystalline), while the other forms an amorphous phase that gives the elastomeric properties. Those block copolymers in which the hard phase is crystalline can be subdivided into four types:

- Polyurethane/elastomer
- Polyester/elastomer
- Polyamide/elastomer
- Polyethylene/poly(α-olefin)

A discussion of the manufacture, structure and properties of these block copolymers is given in Chapter 4. Another type of multi-block copolymer is based on alternating poly(ether-imide) hard segments and polysiloxane elastomer segments [31]. It should be noted that in this case the hard phase is amorphous.

In polyurethane/elastomer block copolymers, the crystalline hard segments are polyurethanes. The amorphous elastomeric segments are either polyethers or polyesters (usually adipates, but polycaprolactones can also be used). The polymers with adipate polyester elastomer segments are tougher and have better resistance to abrasion, swelling by oils, and oxidative degradation. Corresponding polymers with polyether elastomer segments have much better resistance to fungus growth and hydrolytic stability. This is probably the main weakness of the polyester-based materials. Polymers with polyether elastomer segments also are more flexible at low temperatures, more resilient, and thus yield less heat increase under repeated cyclic stress. Polymers with polycaprolactone elastomer segments are premium grades with a combination of superior oil and solvent resistance and quite good hydrolytic stability.

Polyester/elastomer block copolymers share many of the characteristics of polyurethane/elastomer block copolymers. The main difference is that hard segments are crystalline polyesters rather than crystalline polyurethanes. Like the polyurethane/elastomer block copolymers, they are tough, oil resistant materials with high tear strength. They have excellent resistance to compression set and flex cracking. At low strain levels, they show very little hysteresis, i.e., they behave almost like a perfect spring. They are generally somewhat harder than corresponding polyurethane/elastomer block copolymers. Compared with polyurethane/elastomer block copolymers of the same hardness however, they have higher modulus i.e., they are stiffer, better resistance to creep, and show less variation of modulus with temperature. As a result, polyester/elastomer block copolymers claimed to have a wider service temperature range.

Some work has been reported on the use of plasticizers in these materials [32]. Those tested were suitable for use with PVC and were most compatible with the softer grades of polyester/elastomer block copolymers. At least 50 parts of some of these plasticizers could be added to 100 parts of polyester/elastomer block copolymer. The products were softer and more processible. There are also some commercial grades of flame resistant polyester/elastomer block copolymers currently available.

Polyamide/elastomer block copolymers are a newer development in the field of thermoplastic elastomers. They are also block copolymers with alternating hard and soft segments. The hard crystalline segments are polyamides, while the soft segments are either polyethers or polyester. One notable feature is that there are many different polyamides from which to choose, including nylon-6, nylon-6,6, nylon-11, nylon-12, and polyamides containing aromatic groups. These all have different melting points and degrees of crystallinity and so give a wide range of properties. When these variables are combined with different polyethers and polyesters, the range of properties becomes even wider. Quite soft materials can be produced (as low as 65 Shore A hardness), as can harder counterparts (up to 70 Shore D). Those with polyester soft segments have excellent resistance to thermal and oxidative degradation, while those with polyether soft segments are more resistant to hydrolysis and more flexible at lower temperatures. Moisture uptake is normally quite low, but hydrophillic products that are very permeable to water vapor have been described [33]. As far as resistance to deformation is concerned, the upper service temperature depends on the choice of polyamide segment and can be as high as 200 °C. Resistance to oils and solvents is also good.

Polyethylene/poly(α-olefin) block copolymers are newer, lower cost materials. They are very flexible at low temperatures but their upper service temperature is rather low. They have some resistance to oils and solvents and are often blended with other polymers.

Poly(ether-imide)/polysiloxane alternating block copolymers [31] are hard (about 70 on the Shore D scale) and differ from the previous four types in that the hard phase is amorphous. Thus, they do not have as much resistance to oils and solvents. However, these materials have excellent thermal and hydrolytic stability, which justifies their high price.

7.2.1 Replacements for Vulcanized Rubber

This is the most important end use of the polyurethane/elastomer, polyester/elastomer, and polyamide/elastomer block copolymers. They are fabricated into the final article by typical processing techniques developed for conventional thermoplastics, i.e., injection molding, blow molding, extrusion, etc. Their crystalline hard segments make them insoluble in most liquids. Thus, they are often used as replacements for oil-resistant rubbers such as neoprene because they have better physical properties (e.g., tear and tensile strengths) at temperatures up to about 150 °C. Applications include flexible couplings, seal rings, automotive steering boots, gears, footwear (including ski boots), wheels, timing and drive belts, tire chains, industrial hose, and outer coverings for wire and for optical fiber cables. The polyurethane/ elastomer block copolymers with polyether elastomer segments have significant medical applications, including their use in body implants [34]. Polyester/elastomer block copolymers have been used to make transparent replacements for glass bottles for use in medical applications [35]. All three types of these elastomers are usually used uncompounded, although calcium carbonates and radiopaque fillers can be added [36]. The final articles also can be metallized or painted. Special elastomeric paints have been developed which match the appearance of painted automotive sheet metal. Parts painted in this way are used in car bodies. After processing has been completed, the finished parts may be annealed by overnight heating at about 115 °C. This completes the crystallization of the hard phase which in turn, improves the physical properties of the product.

It is most important that all these block copolymers are dried before processing. If they are not, water is released when the hot, molten polymer is processed and the final article contains bubbles and may degrade. Drying times from onc to two hours at temperatures of about 100 °C are recommended and reground polymer or color concentrates should also be dried. During processing, hot, dry air should be circulated through the hopper in which the dry polymer is held to prevent absorbtion of water.

For injection molding applications, mold release can be a problem, as it is with most thermoplastic elastomers. The possible solutions are covered in the section dealing with applications for styrenic block copolymers (see Section 7.1.1). Recommended conditions for extrusion include the use of relatively high compression screws (about 3:1 compression ratio) with length to diameter ratios of at least 24:1. Profile, blown film, and sheet extrusion can all be used with these polymers. They can also be laminated to fabric backings.

In both the molding and extrusion of polyurethane/elastomer block copolymers, the melt temperature should be about 200 °C. Low screw speeds should be used to avoid degradation and the machine should be purged before shutdown [37].

The processing of polyester/elastomer block copolymers is generally easier than polyurethane/elastomer block copolymers because thermal degradation is less of a problem and so the polyester/elastomer block copolymers have a wider processing window. The recommended processing temperature of polyester/elastomer block copolymers varies with the hardness of the material [38]. Thus, the melt temperature during processing of softer (Shore D 40) grades should be about 185 °C while for harder (Shore D 72) grades it should be about 240 °C.

The recommended processing temperatures for polyamide/elastomer block copolymers are about 60 °C above the melting points of the polyamide segments [39] and processing techniques are similar to those used with other polyamides, such as Nylon-6 (poly(ε-caprolactam)).

Suggested applications of polyethylene/poly(α-olefin) block copolymers include molded and extruded goods, such as shoe soles, weatherstrips, tubing, and wire insulation. They can be compounded with fillers and with both naphthenic and paraffinic oils. Processing is similar to that of polyethylene and thermal degradation is not usually a problem.

The main application of the poly(ether-imide)/polysiloxane block copolymers is in flame-resistant wire and cable covering [31], where they combine very low flammability with a low level of toxic products in the smoke. This unusual and vital combination of properties justifies their relatively high price of about $20/lb.

7.2.2 Adhesives, Sealants, and Coatings

This is a relatively small application for these polymers. They are usually applied as hot melts [40], although some grades can be applied from solutions in such polar solvents as methyl ethyl ketone or dimethyl formamide. They have some uses in footwear, including attaching the shoe soles to the uppers, and also in coextrusion, where they act as adhesive interlayers between dissimilar polymers.

7.2.3 Polymer Blends

Polyurethane/elastomer block copolymers have been blended with a wide range of other thermoplastics and elastomers [41] (see also Section 4.2). Non-polar polymers, such as polyolefins, were about the only materials tested that showed little or no compatibility. Probably the most important blending material is plasticized PVC (see Section 5.1), although other elastomers (including styrenic block copolymers) or polar thermoplastics such as polycarbonates can also be used. Blends of plasticized PVC with polyurethane/elastomer block copolymers have good flex life, oil resistance, abrasion resistance, and low temperature flexibility and are used in footwear.

Like S-EB-S block copolymers, polyurethane/elastomer block copolymers can be blended with reactive silicone rubbers to give products intended for medical applications [25]. In fact, they are probably the most important thermoplastic elastomer in this end use.

Polyester/elastomer block copolymers can also be blended with PVC [42]. Both plasticized and unplasticized PVCs have been used and also chlorinated PVC (CPVC) [43]. Compared to the plasticized PVC alone, the blends have better flexibility, particularly at low temperatures, and improved abrasion resistance. In unplasticized PVC and CPVC, the polyester/elastomer block copolymers act as polymeric plasticizers. They can also be blended with thermoplastic polyesters, such as poly(butylene-terephthalate), to give relatively hard, impact resistant products.

Polyamide/elastomer block copolymers can be blended with a number of other polymers, including polystyrene, nylon, polyacrylates, nitrile rubber, and polycarbonates [44].

Polyethylene/poly(α-olefin) block copolymers are recommenced for blending with polypropylene or polyethylene to give hard polymer/elastomer combinations, similar to those described in Section 5.1. The products are claimed to have better properties than the corresponding polypropylene/EPDM blends.

7.3 Hard Polymer/Elastomer Combinations

A very large number of these hard polymer/elastomer combinations have been investigated [45] and are described in more detail in Chapter 5. Morphology is a critical factor and in most cases, a continuous inter-penetrating network of the two components is formed, rather than a dispersion of one in the other [46 to 48]. In some cases, crosslinking agents are added so that the elastomer phase is vulcanized under intensive shear to give a fine dispersion of the crosslinked elastomer in the hard polymer. This is the process known as "dynamic vulcanization" [49] described in Section 5.2. It gives the products better resistance to solvents and to compression set. Combinations (both mixtures and dynamic vulcanizates) of polypropylene with either EPDM or EPR were the first and are still the largest commercial application for these products. They have been available for about 20 years and are believed to have the second largest share of the market for thermoplastic elastomers, after the styrenic block copolymers [50].

The rubber phase can be extended by the addition of mineral oils, giving the products improved processability [48]. However, the amount that can be added is limited by the fact that the rubber phase is already relatively weak, especially if it is not vulcanized, and so further significant loss of strength cannot be tolerated. High density polyethylene improves impact resistance at low temperatures [51]. Judging from values given for their densities [51 to 53], most products are not highly filled, although some grades intended for sound deadening applications are claimed to contain up to 65% by weight of filler [36]. Dynamic vulcanizates based on blends of polypropylene with butyl rubber, natural rubber and nitrile rubber are described in Section 5.2.

When suitably pigmented and stabilized, the hard polymer/elastomer combinations based on polypropylene and either EPR, EPDM, natural rubber, or butyl rubber have good resistance to oxidative and hydrolytic degradation and to weathering. The products can vary quite widely in hardness, from 60 Shore A up to 55 Shore D.

Other commercially important types of hard polymer/elastomer combinations include those based on halogenated polyolefins with ethylene interpolymers (claimed to be single-phase systems) [42, 54] and combinations of PVC with nitrile rubbers [42, 57, 58]. Further details of both types are given in Section 5.1.

Types of nitrile rubber have been developed that are specifically intended for blending with PVC. If the final product is intended for extrusion applications, a pre-crosslinked nitrile rubber (i.e., one crosslinked during its manufacture) is often preferred, but if it is intended for injection molding, uncrosslinked versions may give better results [55, 56]. In some cases, the nitrile rubber is dynamically vulcanized [57]. Published compositions often contain fillers and a liquid ester plasticizer such as dioctyl phthalate [55 to 58]. All these thermoplastic elastomers based on halogenated polylefins or PVC as the hard polymer have good oil resistance, although this resistance is reduced if the rubber phase contains too much plasticizer [56]. The degree of oil resistance is also affected by the type of plasticizer used. The combinations based on dynamically vulcanized nitrile rubbers are claimed to have especially good oil resistance [52, 59].

7.3.1 Hard Polymer/Elastomer Combinations as Replacements for Vulcanized Rubber

This is by far the most significant end use for these materials. Much effort has been devoted to developing specialized grades for various applications, including flame retardant products. They have numerous applications, particularly in the automobile and appliance industry, in weather stripping, and in wire insulation. One significant advantage is the opportunities that they give for part consolidation [60], that is, they allow assemblies of several pieces of rubber and plastic, often held together with metal clamps, to be replaced by a single molded part. As with the styrenic block copolymers (Section 3.1), if a compound contains a particular homopolymer, it adheres well when insert-molded or extruded against the same pure homopolymer, or against other compounds containing it. Thus, blends or dynamic vulcanizates based on a polypropylene hard phase adhere well to polypropylene. Similarly, blends or dynamic vulcanizates based on a PVC hard phase adhere well to PVC.

Polypropylene/EPDM or EPR combinations are especially useful as replacements for general purpose vulcanized rubbers such as natural rubber, SBR, and EPDM. The products are generally pre-compounded by the manufacturer to meet the requirements of the end user. Processing conditions are similar to those used for polypropylene and because the products are relatively stable, degradation during processing is rarely a problem. Scrap can easily be reground and reworked (see the suggestions for grinding soft products previously given in Section 7.1.1). Pre-drying is often advantageous [52]. If the products are to be painted, an undercoat is normally needed, although some directly paintable grades have been developed. Ejection of injection molded parts can be a problem, particularly with the softer grades (again, see the suggestions given in Section 7.1.1). Coextrusion with polypropylene allows the production of profile extrusions with regions of different hardnesses [60], which are used in weather-strips, gaskets, and similar applications.

Polypropylene/natural rubber combinations will probably compete in the lower cost end of the market, while polypropylene/butyl rubber combinations will be especially useful in applications where low gas permeability is important [59].

The materials based on the more polar elastomers, i.e., the combinations of either polypropylene or PVC with nitrile rubbers or the combinations of halogenated polyolefins with ethylene interpolymers, have many applications, especially in areas that take advantage of their excellent oil resistance. All the usual thermoplastic processing techniques (injection molding, blow molding, profile and sheet extrusion) can be used to make end products. Some examples are wire insulation, particularly under the hood, and fuel and hydraulic hose [42, 54, 59]. Simple blends of PVC, nitrile rubbers and DOP should compete effectively in low cost applications, particularly if oil resistance is important. One such application is in injection molded shoe soles. These blends are also reported to have made considerable inroads in automotive applications, especially in Japan.

7.3.2 Polymer Blends

This is not a large application for these thermoplastic elastomers although the use of PP/EPDM or EPR combinations to upgrade scrap polypropylene has been suggested [1].

References

1 *Modern Plastics* (1983) *60*(12), 42
2 Holden, G. and Sun, X.-Y., paper presented at the 4th International Conference on Thermoplastic Elastomer Markets and Products sponsored by Schotland Business Research, Orlando, FL, February 13–15, 1991.
3 Technical Bulletin SC:445-96 (1996) Shell Chemical Co., Houston
4 Technical Bulletin SC:1105-90 (1990) Shell Chemical Co., Houston
5 Buddenhagen, D.A., Legge, N. R. and Zscheuschler, G. (1992 and 1995) U. S. Patents 5,112,900, and 5,407,715 (to Tactyl Technologies, Inc.)
6 Technical Bulletin SC:198-92 (1992) Shell Chemical Co., Houston TX
7 St. Clair, D. J. (1982) *Rubber Chem. Technol. 55*, 208
8 Technical Bulletin SC:72-85 (1985) Shell Chemical Company, Houston
9 Kraus, G., Rollman, K. W. and Gray, R. A. (1979) *J. Adhesion 10*, 221
10 Chu, S. G. and Class, J. (1985) *J. Appl. Poly. Sci. 30*, 805
11 Halper, W. M. and Holden, G. in Chapter 2, *Handbook of Thermoplastic Elastomers*, 2nd Ed., B. M. Walker and C. P. Rader, Eds, (1998), Van Nostrand Reinhold, New York
12 Erikson, J. R. (1990) U. S. Patent 4,904,731 (to Shell Oil Co.)
13 Holden, G. and Chin, S. S., paper presented at the Adhesives and Sealants Conference, Washington D.C., March 1986
14 Mitchel, D. M. and Sabia, R. (1980) *Proceeding of the 29th International Wire and Cable Symposium*
15 Technical Bulletin SC:1102-89 (1989) Shell Chemical Co., Houston
16 Chen, J. Y. (1983, 1993 and 1996), U. S. Patents 4,369,284, 5,262,468, and 5,508,344 (to Applied Elastomerics, Inc.)
17 Bull, A. L. and Holden, G. (1977) *J. Elastomers and Plastics 9*, 281
18 Technical Bulletin SC:165-93 (1993) Shell Chemical Co., Houston
19 Holden, G. and Hansen, D. R. (1990), U. S. Patent 4,904,731 (to Shell Oil Co.)
20 Haaf, W. R. (1990), U. S. Patent 4,167,507 (to General Electric Co.)
21 Gilmore, D. W. and Modic, M. J. (1989) *Plastics Engineering 45(4)*, 51. Reprinted as Technical Bulletin SC:1114-89, (1992) Shell Chemical Co., Houston
22 Holden, G. (1982) *J. Elastomers and Plastics 14*, 148
23 Willis, C. L., Halper, W. M. and Handlin, D. L. Jr., (1984) *Polym.-Plast. Technol. Eng.*, *23*(2), 207
24 Technical Bulletin SC:1216-91 (1991) Shell Chemical Co., Houston
25a Arkles, B. C. (1983) *Medical Device and Diagnostic Industry*, *5*(11), 66
25b Arkles, B. C. (1985) U. S. Patent 4,500,688, (to Petrarch Systems).
26 Van Beem, E. J. and Brasser, P. (1973) *J. Inst. Petroleum* 59, 91
27 Piazza, S., Arcozzi, A. and Verga, C. (1980) *Rubber Chem. Technol. 53*, 994
28 Goodrich, J. L. (1988) "Asphalt and Polymer-Modified Ashphalt Properties Related to the Performance of Asphalt Concrete Mixes", *Asphalt Paving Technol. Proc. AAPT 57*
29 Bouldin, M. G., Collins, J. H. and Berker, A. (1991) *Rubber Chem. Technol. 64*, 577
30 Technical Bulletin SC:1043-90 (1990) Shell Chemical Co., Houston
31 Mihalich, J., paper presented at the 2nd International Conference on Thermoplastic Elastomer Markets and Products sponsored by Schotland Business Research, Orlando, FL, March 15–17, 1989.
32 Hytrel Bulletin HYT-302(R1), (1981) E. I. Du Pont de Nemours & Co.
33 Davis, D. G. and Conkey, J. B., paper presented at the 6th International Conference on Thermoplastic Elastomer Markets and Products sponsored by Schotland Business Research, Orlando, FL, January 15–17, 1992
34 Szycher, M., Poirier, V.L. and Demsey, D. (1983) *Elastomerics 115*(30), 11

35 School, R. in Chapter 9 of Reference 11
36 School, R. (May 7, 1984) *Rubber Plast. News, 49*
37 Ma, E. C. in Chapter 7 of Reference 11
38 Sheridan, T. W. in Chapter 6 of Reference 11
39 Deleens, G. in Reference in Chapter 8B *Thermoplastic Elastomers – A Comprehensive Review*, N. R. Legge, G. Holden and H. E. Schroeder, Eds., (1987) Hanser Publishers, Munich, Vienna, New York.
40 Quinn, F. A., Kapasi, V. and Mattern, R. paper presented at the 6th International Conference on Thermoplastic Elastomer Markets and Products sponsored by Schotland Business Research, Orlando, FL, January 15–17, 1992.
41 Bonk, H. W., Drzal, R., Georgacopoulos, C. and Shah, T. M. paper presented at the 43rd Annual Technical Conference of the Society of Plastics Engineers, Washington, DC, 1985.
42 Hoffman, G. H. in Chapter 6 *Thermoplastic Elastomers*, 2nd Ed., G. Holden, N. R. Legg, R. P. Quirk and H. E. Schroeder Eds., (1996), Hanser Publishers, Munich, Vienna, New York
43 Crawford, R. W. and Witsiepe, W. J. (1973) U. S. Patent 3,718,715 (to Du Pont)
44 Nelb, R. G, and Chen A. T. in Chapter 9 of Reference 42
45 Coran, A. Y., Patel, R. and Williams, D. (1982) *Rubber Chem. Technol. 55*, 116
46 Kresge, E. N. in Chapter 5 of Reference 42
47 Kresge, E. N. (1997) *Rubber World, 217(1)* 30
48 Kresge, E. N. (1991) *Rubber Chem. Technol. 64*, 469
49 Gessler, A. M. (1962) U. S. Patent 3,037,954 (to Esso Research and Engineering Co.)
50 Reisch, M. S. (1996) *Chemical and Engineering News, 74(32)*, 10
51 Shedd, C. D. in Chapter 3 of Reference 11
52 Rader, C. P. in Chapter 4 of Reference 11
53 Holden, G. in Chapter 16 of Reference 42
54 Wallace, J. G. Chapter 5 of Reference 11
55 Kliever, L. B. and DeMarco, R., paper presented at the Symposium on Thermoplastic Elastomers sponsored by the ACS Rubber Division, Nashville, TN November 3–6, 1992
56 Kliever, L. B., paper presented at the 7th International Conference on Thermoplastic Elastomer Markets and Products sponsored by Schotland Business Research, Orlando, FL, February 11–12, 1993.
57 Tandon, P. and Stockdale, M., paper presented at the 4th International Conference on Thermoplastic Elastomer Markets and Products sponsored by Schotland Business Research, Orlando, FL, February 13–15, 1991
58 Stockdale, M., paper presented at the Symposium on Thermoplastic Elastomers sponsored by the ACS Rubber Division, Cincinnati, OH, October 18–21, 1988
59 Thompson, M. J., paper presented at the 6th International Conference on Thermoplastic Elastomer Markets and Products sponsored by Schotland Business Research, Orlando, FL, January 15–17, 1992
60 Mattix, R. B., paper presented at the Symposium on Thermoplastic Elastomers sponsored by the ACS Rubber Division, Nashville, TN November 3–6, 1992.

8 Economic Aspects, Tradenames and Glossary, Future Developments

8.1 Economic Aspects

8.1.1 Price

Economic aspects of thermoplastic elastomers are not simply a function of their price. If they were, thermoplastic elastomers would have achieved little commercial success because their raw material cost is significantly above that of conventional vulcanized rubbers. Equally (and perhaps more) important are the cost savings they bring because of fast processing, recyclability of scrap, and other factors. A significant addition is the new processing techniques that they have introduced to the rubber industry to utilize; including blow molding, comolding and coextrusion, hot melt coating of pressure sensitive adhesives, and direct injection molding of footwear. Thus, rather than a simple cost comparison based on raw material prices, the "value in use" of thermoplastic elastomers must be considered when they are evaluated as possible replacements for more conventional materials. Bearing this caveat in mind, the price ranges of the various types of thermoplastic elastomers are summarized in Table 8.1. This table also includes data on the specific gravity and hardness of the products. The former is particularly important in conjunction with the price. The multiple of the two gives the relative price per unit of volume, and this (rather than the price per unit of mass) is the quantity that determines the cost of the final product.

8.1.2 Commercial Sales

Of course, all the developments discussed in this book would be of much less importance if it were not for the widespread commercial uses of

Table 8.1 Approximate Price and Property Ranges for Thermoplastic Elastomers[a]

	Price Range (Cents/lb.)	Specific Gravity	Hardness
Styrenic Block Copolymers			
S-B-S (Pure)	85–130	0.94	65A–75A
S-I-S (Pure)	100–130	0.92	32A–37A
S-EB-S (Pure)	185–280	0.91	65A–75A
S-B-S (Compounds)	90–150	0.9–1.1	40A–45D
S-EB-S (Compounds)	125–225	0.9–1.2	5A–60D
Polyurethane/Elastomer Block Copolymers	225–375	1.05–1.25	70A[b]–75D
Polyester/Elastomer Block Copolymers	275–375	1.15–1.40	35D–80D
Polyamide/Elastomer Block Copolymers	450–550	1.0–1.15	60A–65D
Polyethylene/Poly(α-olefin) Block Copolymers	80–110	0.85–0.90	65A–85A
Polypropylene/EPDM or EPR Blends	80–120	0.9–1.0	60A–65D
Polypropylene/EPDM Dynamic Vulcanizates	165–300	0.95–1.0	35A–50D
Polypropylene/Butyl Rubber Dynamic Vulcanizates	210–360	0.95–1.0	50A–80D
Polypropylene/Natural Rubber Dynamic Vulcanizates	140–160	1.0–1.05	60A–45D
Polypropylene/Nitrile Rubber Dynamic Vulcanizates	200–250	1.0–1.1	70A–50D
PVC/Nitrile/DOP Rubber Blends	130–150	1.20–1.33	50A–90A
Halogenated Polyolefin/Ethylene Interpolymer Blends	225–275	1.10–1.25	50A–80A

[a]These price and property ranges do not include fire retardant grades or highly filled materials for sound deadening.
[b]As low as 60A when plasticized.

thermoplastic elastomers. We can assess their economic importance by asking how much was (or will be) sold and used. Table 8.2 attempts to answer this question. It shows that over the ten-year period from 1990 to 2000, annual worldwide production is expected to double. Over this decade, production is predicted to grow from almost 1.4 billion lbs to almost 2.9 billion lbs. This is equivalent to an annual growth rate of 7%. As noted above, worldwide sales in 2000 should be almost 2.9 billion lbs. At the current average price of about $1.50/lb., this is equivalent to about $4.3 billion.

8.2 Tradenames and Glossary

Because of this growth and the many applications described in the previous chapter, there are about 50 manufacturers of different types of thermo-

Table 8.2 Estimated Worldwide Consumption of Thermoplastic Elastomers (Millions of pounds per year)

	1990[a]	1995[b]	2000[b]
Styrenic Block Copolymers	625	1070	1435
Polyurethane or Polyester/Elastomer Block Copolymers	245	340	460
Polyolefin Blends and Thermoplastic Vulcanizates	390	580	830
All Others	120	150	170
Totals	1380	2140	2870

[a]Reference [1]
[b]Reference [2]

plastic elastomers. Each of them has at least one tradename for its product(s) and some have several. Many of these tradenames are listed in Tables 8.3 (Styrenic Block Copolymers), 8.4 (Other Block Copolymers) and 8.5 (Hard Polymer/Elastomer Combinations). These tables attempt to describe a constantly changing situation. New manufacturers enter the field and existing businesses are combined or sold. For example, as this book was being written, the Rogal Dutch/Shell Group (probably the largest single manufacturer and a pioneer in the field) announced their intention of selling their entire thermoplastic elastomers business [3].

The elastomers industry, like most others, uses abbreviations and acronyms. A list of some of those used in this book is given in Table 8.6.

8.3 Future Developments

This book has attempted to cover a subject that, over the last 35 years, has grown from small beginnings into a major scientific and business development. Yet this is not necessarily the end of the story. There are many possible structures for new thermoplastic elastomers. We now know much more about how to design new products. We know that the essential requirement is for at least two interdispersed polymeric phases. At normal operating temperatures, one must be fluid (above its T_g) and the other solid (below its T_g or T_m). Additionally, there must be some chemical links or other interactions between them. With this knowledge, polymer (or polymer segments) can be selected or developed to achieve the desired properties.

Currently several trends appear. One is the development of softer products, especially among higher priced, premium thermoplastic elasto-

Table 8.3 Some Tradenames of Thermoplastic Elastomers Based on Styrenic Block Copolymers

Tradename (Mfr.)	Type	Elastomer Segment E	Notes
KRATON D and CARIFLEX (Shell)	Linear and branched	B or I	General purpose, soluble. Also compounded products
VECTOR (Dexco)[a]	Linear	B or I	
SOLPRENE[b] (Phillips)	Branched	B	
TAIPOL (Taiwan Synthetic Rubber Company)	Linear and Branched	B or I	
QUINTAC (Nippon Zeon)	Linear	I	
FINAPRENE (Fina)	Linear	B	General purpose, soluble. Not available as compounded products.
COPERBO (Petroflex)	Linear	B	
TUFPRENE & ASAPRENE (Asahi)	Linear	B	
CALPRENE (Repsol)	Linear and branched	B	
EUROPRENE SOL T (Enichem)	Linear and branched	B or I	
STEARON (Firestone)	Linear	B	High polystyrene content
K-RESIN (Phillips)	Branched	B	Very high polystyrene content. Hard and rigid.
KRATON G (Shell)	Linear	EB or EP	Improved stability. Soluble when uncompounded
SEPTON (Kuraray)	Linear	EB or EP	
DYNAFLEX (GLS)	Linear	B or EB	
MULTI-FLEX (Multibase)	Linear	EB	
HERCUPRENE[c] (J-VON)	Linear	B or EB	Only compounded products
FLEXPRENE (Teknor Apex)	Linear	B	
TEKRON (Teknor Apex)	Linear	EB	
ELEXAR[d]) (Teknor Apex)	Linear	EB	Wire and Cable compounds
C-FLEX (Concept)[e]	Linear	EB	Medical applications Contains silicone oil

[a] Joint venture of Dow and Exxon
[b] No longer made in U.S.A. Similar products are produced by Taiwan Synthetic Rubber Company
[c] Formerly J-PLAST
[d] Formerly produced by Shell
[e] Now Consolidated Polymer Technologies Inc.

Table 8.4 Some Tradenames of Thermoplastic Elastomers Based on Other Block Copolymers

Tradename (Mfr.)	Hard Segment *H*	Elastomer Segment *E*	Notes
ESTANE (B.F. Goodrich) MORTHANE[a] ([b]Rohm and Haas) PELLETHANE[a] (Dow) ELASTOLLAN (BASF) DESMOPAN and TEXIN (Bayer)[c]	Polyurethane	Polyether or amorphous Polyester	Hard and tough. Abrasion and oil resistant. Good tear strength. Fairly high priced
HYTREL (DuPont) LOMOD (GE) URAFIL (Akzo) ECDEL(Eastman) RITEFLEX (Hoechst) ARNITEL (DSM)	Polyester	Polyether	Similar to polyurethanes but more expensive. Better low temperature flexibility. Low hysteresis.
PEBAX (Elf Atochem) VESTAMIDE (Huls) GRILAMID and GRILON (EMS America) MONTAC (Monsanto)[d] OREVAC (Atochem)[d]	Polyamide	Polyether or amorphous polyester	Similar to polyurethanes but can be softer. Expensive. Good low temperature flexibility.
ENGAGE & AFFINITY (Dow) EXACT (Exxon) FLEXOMER (Union Carbide)	Polyethylene	Poly(α-olefins)	Flexible and low cost. Good low temperature flexibility but limited at higher temperatures.
SILTEM (GE)[e]	Poly(ether-imide)	Polysiloxane	Fire retardant, used in wire and cable insulation

[a]Including some with polycaprolactone segments
[b]Formerly Morten International
[c]Formerly marketed by Mobay and Miles
[d]For hot melt adhesives
[e]Amorphous hard segments; all the others are crystalline

Table 8.5 Some Tradenames of Thermoplastic Elastomers Based on Hard Polymer/Elastomer Combinations

Tradename (Mfr.)	Type	Hard Polymer	Elastomer	Notes
REN-FLEX (D&S)[a] HIFAX (Himont) POLYTROPE (Schulmam) TELCAR (Teknor Apex) FERROFLEX (Ferro) FLEXOTHENE (Equistar)[b]	Blend	Polypropylene	EPDM or EPR	Relatively hard, low density, not highly filled
SANTOPRENE (AES)[c] SARLINK 3000 & 4000 (Novacor)[d] UNIPRENE (Teknor Apex) HIFAX MXL (Himont)	DV[f]	Polypropylene	EPDM	Better oil resistance, low compression set, softer
TREFSIN (AES) and SARLINK 2000 (Novacor)[d]	DV	Polypropylene	Butyl Rubber	Low permeability, high damping
VYRAM (AES)	DV	Polypropylene	Natural Rubber	Low Cost
GEOLAST (AES)	DV	Polypropylene	Nitrile Rubber	Oil resistant
ALCRYN (Advanced Polymer Alloys)[e]	Blend	Halogenated Polyolefin	Ethylene Interpolymer	Single phase, oil resistant
SARLINK 1000 (Novacor)[d] CHEMIGUM (Goodyear) APEX N(Teknor Apex)	DV Blend Blend	PVC	Nitrile Rubber	Oil Resistant
RIMPLAST (Petrarch Systems)	Blends of TPEs with Silicone Rubbers			Medical applications

[a]A joint venture between Dexter and Solvay.
[b]Formerly Quantum. Product is a blend of PP and EPR produced in the polymerization reactor.
[c]Advanced Elastomer Systems – a joint venture between Solutia (formerly Monsanto) and Exxon Chemical.
[d]Now a part of DSM.
[e]Formerly DuPont
[f]Dynamic Vulcanizate – a composition in which the soft phase has been dynamically vulcanized, i.e., crosslinked, during mixing.

Table 8.6 Glossary - Some Abbreviations and Acronyms

$\underline{A}$	An amide link
B	A polybutadiene segment
iB	A polyisobutylene segment
CPVC	Chlorinated polyvinylchloride
E	An elastomeric segment, composition unspecified
E	A polyethylene segment
$\underline{E}$	An ester link
EB	A poly(ethylene-co-butylene) segment, typically with a butylene content of about 40%. (more correctly, poly(tetramethylene-co-butylene))
EP	A poly(ethylene-co-propylene) segment
EPR	Ethylene propylene rubber, i.e., poly(ethylene-co-propylene)
EPDM	Ethylene propylene diene rubber , an EPR that also contains a small number of out-of-chain double bonds
DOP	Dioctyl phthalate (more correctly bis (2-ethyl hexyl) phthalate)
H	A hard segment, composition unspecified
HDI	Hexamethylene diisocyanate
HDPE	High density polyethylene
HNBR	Hydrogenated nitrile butadiene rubber
I	A polyisoprene segment
MDI	Diphenylmethane 4,4′ diisocyanate
NBR	Nitrile butadiene rubber, i.e., poly(acrylonitrile-co-butadiene)
S	A polystyrene segment
S*	A substituted polystyrene segment , e.g., poly (α-methyl styrene)
SBR	Styrene butadiene rubber, i.e., poly(styrene-co-butadiene)
P	A polymer, composition unspecified
P	A polypropylene segment
iP	An isotactic polypropylene segment
LLDPE	Linear low density polyethylene
PBT	Poly(butylene terephthalate), more correctly, Poly(tetramethylene terephthalate)
PPE	Poly(phenylene ether), poly(phenylene oxide) more correctly, poly(2,6- dimethyl-1,4-phenylene ether)
PVC	Polyvinylchloride
TDI	2,4 Toluene-diisocyanate
T_g	The glass transition temperature of a polymer or segment
T_m	The crystal melting temperature of a polymer or segment
TPE	Thermoplastic Elastomer
$\underline{U/E}$	A urethane or ester link

mers such as poylurethanes, polyesters, and polyamides. Another is the trend toward thermoplastic elastomers with better stability, both physical and oxidative, at higher temperatures. Flame resistance, oil resistance, and the ability to be comolded or coextruded will all be enhanced. Melt-mixing of polymer blends is an obvious low cost route to new products. New polymerization techniques will allow reproducible production of block copolymers with low cost crystalline hard segments (isotactic polypropylene is an obvious choice) and amorphous soft segments. In many applications, these new products will compete against the entire range of conventional crosslinked rubbers. Of course, there will be limitations. It is doubtful if thermoplastic elastomers will ever enter the largest end use of conventional rubbers, automobile and truck tires. Still, they will expand into almost every other end use. In other words, the scope and utility of thermoplastic elastomers will be much broader than what it is generally considered today. This is an area for research and development with a scope about as broad as we can imagine.

References

1 Holden, G., Legge, N. R., Quirk, R. P. and Schroeder, H. E. in Chapter 17. *Thermoplastic Elastomers*, 2nd Ed., G. Holden, N. R. Legge, R. P. Quirk and H. E. Schroeder Eds., (1996), Hanser Publishers, Munich, Vienna, New York
2 Reisch, M. S. (1996) *Chemical and Engineering News, 74(32)*, 10
3 Layman, P. and Reisch, M. S. (1998) *Chemical and Engineering News, 76(51)*, 6

Subject Index

R

S